AGRICULTURE ISSUES AND POLICIES

U.S. DAIRY SUPPORT PROGRAMS

FARM BILL PROVISIONS AND GROSS MARGIN-DAIRY INSURANCE

AGRICULTURE ISSUES AND POLICIES

Additional books in this series can be found on Nova's website under the Series tab.

Additional e-books in this series can be found on Nova's website under the e-book tab.

AGRICULTURE ISSUES AND POLICIES

U.S. DAIRY SUPPORT PROGRAMS

FARM BILL PROVISIONS AND GROSS MARGIN-DAIRY INSURANCE

CARLA EDIE
EDITOR

New York

For permission to use material from this book please contact us:
Telephone 631-231-7269; Fax 631-231-8175
Web Site: http://www.novapublishers.com

NOTICE TO THE READER

Additional color graphics may be available in the e-book version of this book.

LIBRARY OF CONGRESS CATALOGING-IN-PUBLICATION DATA

ISBN: 978-1-63321-715-7

Published by Nova Science Publishers, Inc. † New York

Contents

PREFACE

Chapter 1 – The 2014 Farm Bill makes significant changes to the structure of U.S. dairy support programs, including the elimination of several major price and income support programs from the 2008 Farm Bill, the extension of several smaller dairy programs, and the addition of two new programs, the Margin Protection Program and the Dairy Product Donation Program.

This report describes the major dairy provisions contained in the 2014 Farm Bill as well as Congressional Budget Office cost projections of historical program outlays compared with outlays under the new dairy programs. The report also includes a discussion of potential issues related to the new dairy policies.

Chapter 2 – Public risk management policies for dairy producers have the potential to induce expansion in milk supplies, which might lower farm-level prices and offset risk-reduction benefits. An evaluation of USDA's Livestock Gross Margin-Dairy (LGM-Dairy) insurance program finds economic downside risk significantly reduced, with potential to induce modest supply expansion (0 to 3 percent) if widely adopted. Supply impacts are likely limited due to relatively low participation levels and a minimal ("inelastic") supply response to risk. LGM-Dairy is more flexible and convenient than other risk management tools, such as hedging directly in futures or options markets, especially for small farms.

In: U.S. Dairy Support Programs
Editor: Carla Edie
ISBN: 978-1-63321-715-7

Chapter 1

DAIRY PROVISIONS IN THE 2014 FARM BILL (P.L. 113-79)*

Randy Schnepf

SUMMARY

The 2014 farm bill (P.L. 113-79), which was signed into law on February 7, 2014, makes significant changes to the structure of U.S. dairy support programs, including the elimination of several major price and income support programs from the 2008 farm bill (P.L. 110-246), the extension of several smaller dairy programs, and the addition of two new programs.

Three of the principal dairy support programs under the 2008 farm bill—the Dairy Product Price Support Program (DPPSP), the Milk Income Loss Contract (MILC) program, and the Dairy Export Incentives Program (DEIP)—are eliminated. These programs are replaced by two new support programs that are authorized for the five-year period of the 2014 farm bill, FY2014-FY2018—the Margin Protection Program (MPP) and the Dairy Product Donation Program (DPDP). However, MILC will be temporarily available until September 1, 2014, or until the implementation date of the new MPP, whichever comes earlier.

---

* This is an edited, reformatted and augmented version of a Congressional Research Service publication R43465, prepared for Members and Committees of Congress dated May 6, 2014.

The MPP is a voluntary program that pays participating farmers an indemnity when a formula-based national margin—referred to as the Actual Dairy Production Margin (ADPM) and calculated as the national average farm price for all milk minus a national-average feed cost ration—falls below a producer-selected insured margin that can range from $4.00 per hundredweight (cwt.) to $8.00/cwt. in $0.50/cwt. increments. Producers must pay an annual administrative fee of $100 for each participating dairy operation. In addition, producers must pay a premium that rises steadily for higher coverage levels starting at the $4.50/cwt. coverage level. The minimum $4.00/cwt. coverage is fully subsidized and has no farmer-paid premium. The premium structure is further divided based on size of operations—lower premiums are charged for the first 4 million pounds (lbs.) of annual milk marketings, higher premiums are charged on marketings above 4 million lbs. As an added incentive to encourage participation by smaller dairy operations (with annual milk marketings under 4 million lbs.), premiums will be reduced by 25% across the board for all coverage levels except the $8.00/cwt. level during calendar 2014 and 2015.

The DPDP requires USDAto procure and distribute certain dairy products when the ADPM falls below $4.00/cwt. for each of the immediately preceding two months. DPDP dairy product distribution is required to target individuals from low-income groups and not be allowed for resale into commercial markets. Purchases and distribution under the DPDP end after three months or if certain other triggers related to the ADPM and U.S.-to-international dairy product prices are met.

Several programs from the 2008 farm bill were extended through FY2018 including the Dairy Forward Pricing Program, the Dairy Indemnity Program, and certain provisions to augment the development of export markets under the National Dairy Promotion and Research Program (i.e., the dairy check off program). In addition, the final bill adopted a provision that requires USDA to adhere to standard rulemaking procedures.

Separately, federal milk marketing orders have permanent statutory authority and continue intact, as does the Livestock Gross Margin for Dairy Cattle program (LGM-D) and the suite of Dairy Import Tariff Rate Quotas (TRQs) that limit access to the U.S. domestic market by lower-priced foreign dairy products. The permanent Dairy Price Support Program contained in the Agricultural Act of 1949 (P.L. 81-439) is suspended but would be reactivated should MPP expire at the end of FY2018 without replacement or extension.

INTRODUCTION

After several years of growing producer discontent with the apparent ineffectiveness of the federal dairy price support program and growing concerns about rapidly escalating dairy feed costs, the U.S. dairy industry advocated for substantial changes in the nature of federal dairy support. As a result, the 2014 farm bill (the Agricultural Act of 2014; P.L. 113-79), which was signed into law on February 7, 2014, makes significant changes to the structure of U.S. dairy support programs including the elimination of several major price and income support program provisions from the 2008 farm bill (the Food, Conservation, and Energy Act of 2008, P.L. 110-246), the extension of several smaller dairy programs, and the addition of two new programs.

This report describes the major dairy provisions contained in the 2014 farm bill as well as Congressional Budget Office (CBO) cost projections of historical program outlays compared with outlays under the new dairy programs.[1] The report also includes a discussion of potential issues related to the new dairy policies. Finally, a table at the end of this report (**Table A-1**) provides a side-by-side comparison of dairy provisions from the 2008 and 2014 farm bills.[2]

Reaching a final compromise on U.S. dairy policy, as contained in the 2014 farm bill, was a long and arduous task involving considerable debate over what the nature and role of federal support programs should be. The following section provides background on the shift of federal support from dairy product prices to the dairy operating margin.

ORIGINS OF THE NEW DAIRY POLICY

Federal dairy price supports were first established by federal law in 1949 and modified in subsequent legislation including the 2008 farm bill (P.L. 110-246), which established the Dairy Product Price Support Program (DPPSP).[3] DPPSP indirectly supported the farm price of fluid milk at $9.90 per hundred lbs. (i.e., hundredweight or cwt.) through government purchases of dairy products from dairy processors at statutorily set prices.[4] The program was countercyclical, in that government purchases occur when product prices are low, and cease as product prices rise above support levels. However, by the mid-1990s, the annual farm price of milk had trended higher, albeit subject to an increasingly volatile pattern (**Figure 1**), whereas the federal support rate was flat at $9.90 per cwt.

Volatile milk prices made planning more difficult and made dairy producers more vulnerable to unexpected or sustained increases in the cost of feed (the major cost component of dairy production). Milk producers have argued since the early 1990s that support levels had become too low, relative to market prices and costs of production, to provide meaningful support. More recently, milk producers argued that support based strictly on the price of milk failed to account for the sharp escalation of feed costs that has occurred since 2006 (**Figure 2**).

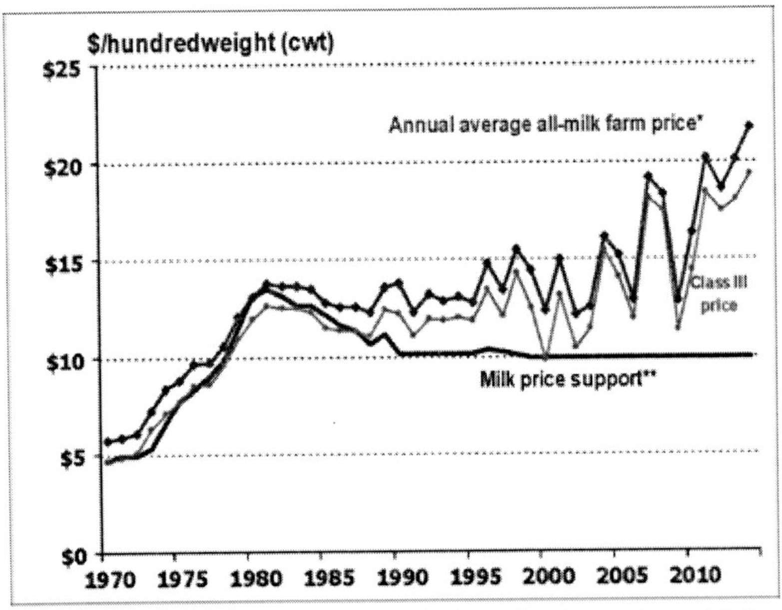

Source: U.S. Dept. of Agriculture; World Agricultural Supply and Demand Estimates (WASDE), March 10, 2014.

Notes: National average price received by farmers, all milk, and the announced Class III price, are USDA data; 2014 prices are USDA forecasts. ** The national price support for milk was statutorily established at $9.90 per cwt. from 1998 until 2008. Beginning in 2008, government purchase prices were established for individual dairy products, but with essentially the same effect as supporting raw milk at $9.90 per cwt. These price supports were repealed by the 2014 farm bill (P.L. 113-79).

Figure 1. Milk Prices Moved Well Above Previous Support Levels By 1990.

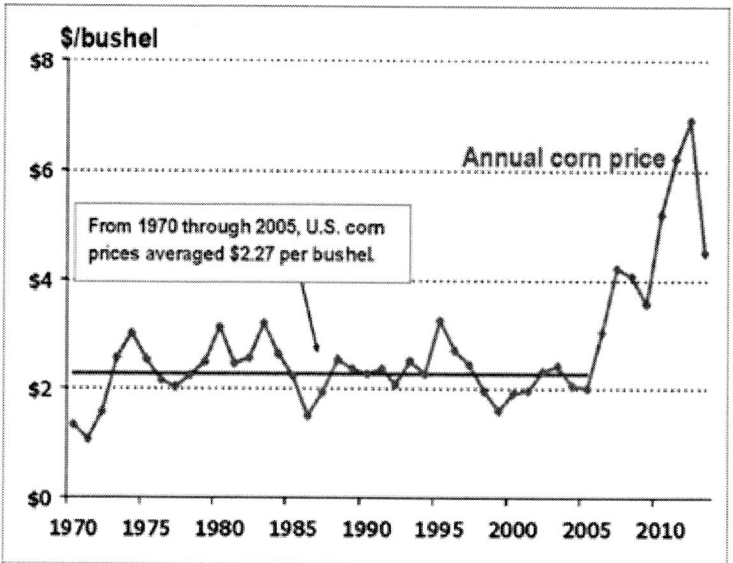

Source: USDA, WASDE, March 10, 2014. The national average price received by farmers for corn for the 2012 crop year averaged a record $6.89 per bushel while the 2013 price is estimated by USDA at $4.50 per bushel.

Notes: Corn is the principal feed grain used in the United States. Prices for other feed grains and hay are closely correlated with the price of corn.

Figure 2. Feed Prices, Led by Corn, Have Risen Sharply Since 2006.

Since the emergence of the U.S. ethanol industry as a major source of corn demand in 2006, U.S. feed grain markets have surged to new price levels that are two to three times above the levels that persisted during the previous four decades. Rising feed costs are of particular concern to dairy producers because they represent a substantial portion of the cost of milk production—in 2011, feed costs accounted for 80% of operating costs and 54% of total costs of milk production (compared with 71% and 37% shares during the 2000-2004 period).[5]

The combination of volatile milk market prices rising well beyond support levels coupled with rapidly rising feed costs contributed to a shifting of focus away from price supports and towards protecting or guaranteeing some portion of the margin between milk prices and feed costs.

This refocusing of policy interest on the dairy operating margin was formalized by a proposed dairy margin protection program published in June 2010 by the National Milk Producers Federation (NMPF) called the

Foundation for the Future (FTF).[6] A version of the FTF was introduced in the 112th Congress as H.R. 3062, The Dairy Security Act (DSA), by House Agriculture Committee Ranking Member Collin Peterson on September 23, 2011. Through the legislative process FTF eventually evolved into the new dairy program described in the following sections. An appendix at the end of this report (**Appendix**) provides additional detail on the legislative debate behind the new dairy policy.

DAIRY PROVISIONS IN THE 2014 FARM BILL

The 2014 farm bill repealed several previously existing dairy programs, extended others, added some new programs, and was silent on several existing dairy programs that derive their authority from outside of the farm bill. Each of these aspects is briefly summarized below.

Programs Repealed

The three major price and income support programs from the 2008 farm bill are eliminated. These include the following.[7]

- The **Dairy Product Price Support Program (DPPSP)** is repealed immediately. DPPSP mandated that USDA directly support the price of cheese, nonfat dry milk, and butter through acquisitions at statutorily set minimum purchase prices in order to support the farm price of milk at $9.90 per hundred lbs. or hundred-weight (cwt.). The program also included USDA resale of acquired dairy products under certain price and stock-volume conditions. Some variation of a dairy price support program has been in effect since the Agricultural Act of 1949 first required USDAto support the farm price of milk.
- The **Milk Income Loss Contract (MILC)** program is temporarily extended until September 1, 2014, or until the implementation date of the new margin program (described below), whichever comes earlier and after which it is permanently repealed. MILC is a counter-cyclical payment program first authorized in the 2002 farm bill (P.L. 107-171) that makes payments available on up to 2.985 million lbs. of milk whenever the monthly Boston Class I price of fluid milk falls below a threshold of $16.94/cwt. adjusted for feed cost changes.

- The **Dairy Export Incentives Program (DEIP)**, originally authorized by the 1985 farm bill (P.L. 99-198), is repealed immediately. DEIP provided cash bonuses to U.S. dairy exporters when certain international dairy market conditions were met.

The **Federal Milk Marketing Order Review Commission**—established by the 2008 farm bill (§1509) to conduct a comprehensive review and evaluation of FMMO and non-FMMO systems—is also repealed effective immediately. The review commission-which was subject to annual funding appropriations—was never funded.

Programs Extended

Certain other programs that were reauthorized by the expired 2008 farm bill (P.L. 110-246) are extended through the life of the 2014 farm bill—that is, FY2018. These include the following.

- The **Dairy Forward Pricing Program** allows farmers to voluntarily enter into forward price contracts with milk handlers for pooled milk used for manufactured products (Classes II, III, and IV) under Federal Milk Marketing Orders (FMMOs). The program allows regulated handlers to pay farmers in accordance with the terms of a forward contract instead of paying the minimum FMMO blend price for pooled milk. The prices paid by milk handlers under the contracts are deemed to satisfy the minimum price requirements of FMMOs. The 2014 farm bill allows for new contracts until September 30, 2018, but no contract can extend beyond September 30, 2021.
- The **Dairy Indemnity Payment Program (DIPP)** makes payments to dairy producers when a public regulatory agency directs them to remove their raw milk from the commercial market because it has been contaminated by pesticides, nuclear radiation or fallout, or toxic substances and chemical residues other than pesticides through no fault of their own. Payments also are made to manufacturers of dairy products, but only for products removed from the market because of pesticide contamination.
- The **Dairy Promotion and Research Program** is a generic dairy product promotion, research, and nutrition education program, funded by a mandatory 15¢/cwt. assessment on milk produced and marketed

in the 48 contiguous states. Importers in all 50 states, the District of Columbia, and Puerto Rico must also pay an assessment rate of 7.5¢/cwt. of the milk equivalent in imported dairy products. USDAissues regulations on the time and method of importer payments.

Continuation of Existing Programs under Non-Farm-Bill Authority

Certain dairy programs derive their operating authority outside of the farm bill and simply continue to operate independent of the enactment of the 2014 farm bill. These include, but are not limited to the following.

- **Federal Milk Marketing Orders (FMMOs)**, which exist under permanent authority, established by federal law in theAgricultural MarketingAgreement Act of 1937, are 10 geographically defined fluid milk marketing areas. The FMMO system regulates milk marketing across state lines but within these 10 explicitly defined and geographically aligned multi-state regions.[8] Within each FMMO, dairy processors or handlers (i.e., milk buyers) are required to pay a minimum price for farm milk depending on its end use.
- **Dairy Import Tariff Rate Quotas (TRQs)** are a system of product-specific import quotas—with low in-quota tariffs and high, often-prohibitive above-quota tariffs—designed to protect higher-priced domestic dairy products by limiting the importation of lower-priced foreign dairy products. Dairy TRQs are part of the Harmonized Tariff Schedule of the United States and are unaffected by changes made in the 2014 farm bill.[9]
- The **Dairy Price Support Program** is authorized under "permanent farm law" contained in the 1949 Agricultural Act, but is suspended by periodic passage of new farm legislation.[10] Under permanent law, USDA is required to purchase manufactured milk and butterfat products in sufficient quantities to raise demand in order to raise the farm price of fluid milk to the equivalency of 75% to 90% of a 1910-1914 parity price index—which would result in milk support prices near $40/cwt. double the average all-milk farm price of $20/cwt. during 2013. Reversion to permanent dairy law has been referred to as the "milk price cliff."

- The **Dairy Product Mandatory Reporting Program** requires manufacturers to report to USDA the price, quantity, and moisture content of dairy products sold. Quarterly audits are to be undertaken to ensure compatibility between submitted information and related dairy market statistics.
- The **Livestock Gross Margin Insurance for Dairy (LGM-D)** is an ongoing pilot program available for purchase from private insurers through USDA's permanently authorized federal crop insurance program. LGM-D provides protection to dairy producers when feed costs rise or milk prices drop. Gross margin is the market value of milk minus feed costs. LGM-D uses futures prices for corn, soybean meal, and milk to determine the expected gross margin and the actual gross margin.[11]
- The **Fluid Milk Processor Promotion Program**, established by the 1990 farm bill (P.L. 101-624) with subsequent reauthorizations, develops and finances generic advertising programs designed to maintain and expand markets and uses for fluid milk products produced in the contiguous 48 states and the District of Columbia. The program is funded through a 20¢/cwt. assessment paid by processors on all milk processed for fluid consumption. The fluid milk order was approved by a referendum among fluid milk processors and became effective December 10, 1993. The program originally required periodic congressional reauthorization; however, the 2002 farm bill gave it permanent authority.

Revival of a Previously Expired Provision

The 2014 farm bill also included a provision (§1410(d)) that removed the expiration dates in U.S. code (7 U.S.C. 7235[a][2]) related to the petition and approval by California dairy producers to designate the state of California as a separate Federal Milk Marketing Order (FMMO) with the potential to use a Class I quota plan to determine individual producer payments. California currently operates its dairy industry under a state marketing order, but in recent years has been considering adopting an FMMO in order to alter its current milk class pricing structure. By removing the expiration dates, the 2014 farm bill actively revives the FMMO option for California's dairy industry.

USDA's Agricultural Marketing Service would still have to receive a formal petition from California's producers and conduct a hearing to collect evidence about the market and hear testimony on desired provisions before California's dairy producers actually decide whether to adopt—via referendum—an FMMO.

New Programs

The 2014 farm bill replaced the repealed price support programs—DPPSP, MILC, and DEIP— with two new support programs that are authorized for the five-year period of the 2014 farm bill, FY2014-FY2018—the Margin Protection Program (MPP) and the Dairy Product Donation Program (DPDP). These two programs are described in the following two sections.

Although the dairy provisions of the 2014 farm bill provide important structure and direction concerning the application of the new programs, substantial detail still needs to be worked out by USDA in regulations in order to implement the new programs. As a result, this report is "preliminary" in the sense that USDA implementing regulations have not yet been proposed. Final regulations are expected sometime before September 1, 2014, at which point the new dairy programs are required to become operational.

The Margin Protection Program (MPP) for Dairy Producers

MPP Implementation Specifics

USDA is mandated to establish a new margin protection program (MPP) for dairy producers no later than September 1, 2014.[12] MPP is a voluntary program intended to provide milk producers with protection from low operating margins.

Defining the Margin

A key aspect of the MPP is creating a timely and transparent measure of an average dairy-production operating margin that will be useful across all dairy production regions. For MPP purposes, the operating margin is defined as the difference between the farm-level price of a hundred lbs. of milk—referred to as hundredweight or cwt. of milk—and the cost of an average feed ration necessary to produce a cwt. of milk. To construct this margin, MPP will use USDA-reported monthly national average price data for all classes of milk (called the all-milk price) and for the three principal feeds that represent the bulk of purchased feeds in dairy rations (corn, soybean meal, and alfalfa hay) to construct an estimate of the operating margin (see box below).

MPP Eligibility

All U.S. dairy producers are eligible to participate in the margin protection program. If a dairy operation consists of more than one dairy producer, all of the dairy producers shall be treated as a single dairy producer for purposes of participation in the margin protection program. If a dairy producer operates two or more dairy operations, each dairy operation shall separately register to participate in the margin protection program. USDA is explicitly tasked with developing regulations to prevent a dairy producer from reconstituting a dairy operation for the purpose of capturing payments under MPP.

Signing up for Margin Protection

Participating in MPP is voluntary. To participate in MPP for any given year, each dairy operation must sign up and pay an annual registration fee of $100.[16] USDA is charged with specifying the manner and form by which a participating dairy operation may register to participate in the MPP. The 2014 farm bill is silent on an explicit sign up period, but designated that decision to USDA. To this end, USDA will announce a registration (or signup) period in the *Federal Register* including the manner and form of registration (or signup). It seems likely that many of the rules on producer requirements for submitting production, ownership, and other records would be similar if not identical to those used for the MILC programs.

Actual Dairy Production Margin (ADPM) =
Milk Returns over Feed Costs

The ADPM is defined as the difference between the monthly national average "all-milk" farm price and an average, formula-derived monthly value for the cost of a representative dairy feed ration.[13]

Margin per cwt. = (All-Milk Price per cwt.) – (Feed Cost per cwt.)

Weighted Feed Cost Formula

The average price paid for feed used by a nationally representative dairy operation to produce a cwt. of milk is based on price data for the three major feed ingredients—corn, soybean meal, and alfalfa hay. Monthly price data for these three feedstuffs are combined into a weighted feed cost estimate per cwt. of milk production using the following formula.[14]

Feed Cost per cwt. = (1.0728 x corn price) + (0.00735 x soybean meal price) + (0.0137 x alfalfa hay price)

where the corn price is in \$/bushel and the soybean meal and alfalfa hay prices are in \$/ton.

Note: It is important to note that the MPP is calculated as a national average margin. All participating dairy operations will face the same national ADPM estimate, irrespective of their individual circumstances. Thus, the MPP does not guarantee an individual producer's margin. Instead, it is assumed that each producer's margin will vary in a way that is correlated with the national average.[15] This formulation is used, in part, because the data are both transparent and readily available at the national level, thus facilitating its routine and timely calculation, and also because feed costs traditionally are the most variable component of dairy production operating margins.

The component mix of feed rations likely varies substantially across regions and individual dairy operations based on local pasture, feed preferences, and other conditions.

In addition, important milk production costs are necessarily excluded from this formula, including labor, utilities, depreciation, capital, veterinary services, and nutritional supplements. Thus, this operating margin formula is a crude indicator of dairy profitability. To the extent that excluded operating cost items and relevant feed ration vary greatly across individual operations, this variation will likely be addressed by individual producers when determining their desired level of margin coverage.

Producers who sign up for MPP are ineligible to sign up for the Livestock Gross Margin for Dairy Cattle program (LGM-D) offered by USDA's Risk Management Agency (RMA).[17] Dairy producers may, on an annual basis, change their MPP coverage level or opt out entirely—a year can be skipped without jeopardizing or prejudicing enrollment in a future year. Further clarification is expected in USDA's program implementation regulations.

MILK PRODUCTION COVERAGE UNDER MPP

Annual Production History (APH)

At the initial sign up, the annual milk production history (**APH**) of a participating farm is defined as the highest annual milk marketings of the dairy operation during any one of the three calendar years 2011, 2012, or 2013. Special provisions are made for new dairy operations that were not yet in operation for a full year during the 2011 to 2013 period—they may choose between two methods to extrapolate their available production history to a 12-month equivalent. Either they base the extrapolation on the volume of actual milk marketings for the months that the dairy operation has been in operation, or they use an estimate of their annual milk marketings based on the farm's actual milking herd size and the national average yield data published by USDA. In subsequent years (i.e., 2014 and beyond), USDA shall adjust each participating farm's APH to reflect any increase in the national average milk production. This allows an operator's margin protection coverage potential to grow with the national average. However, if an individual farm expands its milk production more rapidly than the national average increase, then that portion of his expanded milk production that is in excess of the national average increase will not be eligible for coverage under MPP.

Producer Choice: Margin Protection Level & Coverage Level

Under MPP, a participating producer must select both the margin protection level which ranges from $4.00/cwt. to $8.00/cwt. in $0.50/cwt. increments, and the coverage level or portion of the farm's APH that the producer wants to protect ranging from 25% to 90% in 5% increments.

The ADPM, APH, margin protection level, and coverage level are used jointly to determine: when an indemnity payment is made; what the payment total will be for a participating dairy operation; and what the annual premium or cost will be for the selected coverage levels (see description below on payments and premiums).

Consecutive Two-Month Periods

For purposes of evaluating when a margin payment has been triggered and, if so, the amount of the payment, the ADPM is calculated in specific two-month periods. Each calendar year is broken into the following two-month periods: January-February, March-April, May-June, July-August, September-October, and November-December.

Note that a low single-month average margin does not trigger a margin protection payment if the two-month average is above the trigger. For example, assume a producer has selected a $6.00 margin threshold. Then a January margin of $5.80/cwt. followed by a February margin of $6.30/cwt. produces a two-month average of $6.05/cwt., which would fail to trigger the margin threshold.

USDA is instructed to determine the ADPM as soon as possible after the necessary prices are reported. NASS full-month price estimates—not preliminary estimates—must be used for both months in calculating the two-month average. As a result, the two-month average margin calculation will not be available until a full month after the two-month period has expired.

One-Month Period Average Margins

Average margins are calculated for one-month periods for purposes of evaluating whether a Dairy Product Distribution Program (DPDP) threshold has been triggered (described in detail below).

DPDP purchase and distributions are triggered when the ADPM average is less than $4.00/cwt. for two consecutive months.

No Cap or Limit on Participation

Unlike the MILC program where payments each fiscal year were limited to the first 2.985 million lbs. of milk for producers with an Adjusted Gross Income (AGI) under $500,000, participation in the MPP has no explicit cap related to a producer'sAGI or the size of a dairy operation—in other words, there is no production or dollar payment limitation associated with the dairy margin program. Instead, margin protection payments would be limited by how much of a producer's historical milk marketings are covered based on an elected coverage percentage level (see box above).

Timing of Payments

The 2014 farm bill does not specify a timetable for payments. Instead, USDA is instructed to develop rules for collecting premiums and paying out indemnities.

Likelihood of an MPP Payment

The historical frequency of the average margin levels (**Table 1**) provides information concerning the likelihood of future payments at different margin thresholds.

Table 1. Margin Distribution, January 2000 through March 2014

Margin Range	Number of Months	Share (%)
Margin < $4.00	12	7.0%
$4.00 < Margin < $6.00	8	4.7%
$6.00 < Margin < $8.00	54	31.6%
Margin > $8.00	97	56.7%
Total	**171**	**100%**

Source: CRS calculations using USDA data and based on the two-month period margins (Jan.-Feb., Mar.-Apr., May-June, July-Aug., Sept.-Oct., and Nov.-Dec.) as described in the box entitled, "Milk Production Coverage Under Margin Protection" where Margin per cwt. = (All-Milk Price per cwt.) – (Feed Cost per cwt.).

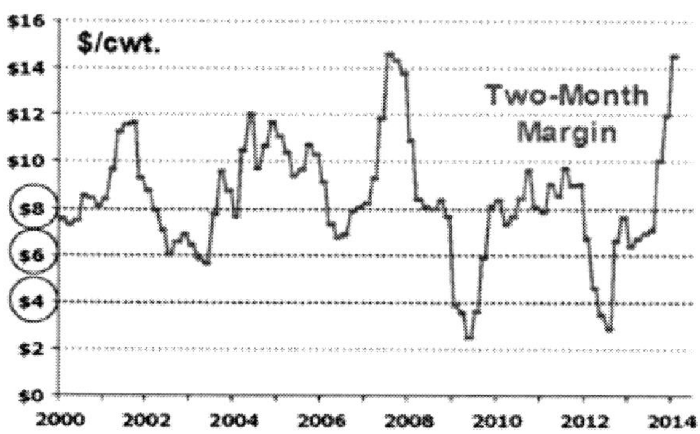

Source: Margin (national average all-milk price minus average cost of feed ration) calculated by CRS using USDA data (*Agricultural Prices*, March 28, 2014) and based on the two-month periods: Jan.-Feb., Mar.-Apr., May-June, July-Aug., Sept.-Oct., and Nov.-Dec.

Figure 3. Historical Dairy Operating Margin: (All-Milk Price) Less (Avg. Feed Cost).

Using ADPM estimates for successive two-month periods since January 2000, the ADPM has been below $4/cwt. in 7% of the two-month periods and above $8/cwt. in nearly 57% of the periods. Margins within the $6/cwt. to $8/cwt. range occurred in nearly one-third of the two-month periods. During this same period (January 2000 to March 2014) the national average ADPM was $8.45/cwt. Thus, the maximum MPP coverage level of $8.00/cwt. represents 95% of the national average ADPM.

See **Figure 3** for a depiction of how often the monthly margin would have fallen below the $8.00/cwt., $6.00/cwt., and $4.00/cwt. thresholds since January 2000.

MPP Payments

Whenever the calculated Actual Dairy Production Margin (ADPM) falls below the selected MPP margin threshold for a consecutive two-month period, a payment will be made on the selected coverage-level portion of a participating producer's APH. The MPP payment rate per cwt. is equal to the difference between the selected MPP threshold and the average ADPM as calculated for that two-month period.

MPP Payment Rate per cwt. = (Selected MPP Threshold) – (2-month average ADPM)

To determine the total MPP payment for a specific two-month period, the MPP payment rate times the coverage percent is applied to the producer's APH, but prorated to a two-month period (i.e., APH divided by six).

MPP Total Payment = (MPP Payment Rate) * (Coverage %) * (APH/6)

MPP payments will continue as long as the ADPM is less than the selected MPP margin threshold for consecutive two-month periods. MPP payments cease when the ADPM reaches or exceeds the selected MPP margin threshold for a two-month period.

EXAMPLE OF MPP PAYMENT CALCULATION

Suppose that for a particular two-month period the average all-milk price is $20.00/cwt. and the formula-determined feed ration per cwt. is $15.50, such that the ADPM is $4.50/cwt. Consider a dairy producer that traditionally has about 500 cows in his operation, but that is slowly expanding. The producer has selected a $6.50/cwt. MPP margin threshold with a 90% coverage level. Assume his APH is 12 million lbs. (or 120,000 cwt.), then the MPP payment for the two-month period will be calculated as follows. The MPP payment rate would be based on the difference between the selected $6.50 margin threshold and the lower ADPM:

MPP Payment Rate per cwt. = $6.50 - $4.50 = $2.00

The MPP payment for the two-month period equals the payment rate times the coverage percentage of 90% times the pro-rated APH (i.e., 120,000 cwt./6 = 20,000 cwt.):

MPP Payment = ($2.00) * (90%) (20,000 cwt.) = $36,000

Note that the MPP payment example is for a specific two-month period and would have to be recalculated for each succeeding two-month period based on any changes in the average margin. MPP payments would continue as long as the two-month ADPM is less than the selected MPP margin level. These two-month payments are in contrast to the annual administrative fee and the MPP premium, which are only paid once in a year.

MPP Premiums

Participating producers must pay premiums that start at the $4.50/cwt. margin protection level and rise with higher coverage levels up to the $8.00/cwt. maximum. No premium is charged for the minimum $4.00/cwt. margin protection. The premium schedule also differentiates based on the size of operation. Asubstantially lower premium schedule is used for the first 4 million lbs. of annual milk marketing (**Table 2**). In addition, a special 25%

discount is available for premiums on the first 4 million pounds of milk marketings during each of calendar 2014 and 2015.

Table 2. MPP Premium Rates per Hundredweight (cwt.) of Milk Coverage All values are in $/cwt

Margin Protection Level	1st 4M lbs. of APH		APH > 4M lbs.
	Discounted Premium[a]	Premium	Premium
$4.00	$0.000	$0.000	$0.000
$4.50	$0.075	$0.010	$0.020
$5.00	$0.01875	$0.025	$0.040
$5.50	$0.030	$0.040	$0.100
$6.00	$0.04125	$0.055	$0.155
$6.50	$0.0675	$0.090	$0.290
$7.00	$0.1625	$0.217	$0.830
$7.50	$0.225	$0.300	$1.060
$8.00	$0.475	$0.475	$1.360

Source: 2014 farm bill (P.L. 113-79; §1407).

Note: M = million; APH = Annual Production History.

a. Premiums on the first 4 million lbs. are reduced by 25% during each of calendar years 2014 and 2015.

The premium rate differential based onAPH is significant because, in 2011, approximately 88% of U.S. dairy farms had annual milk production of 4 million lbs. or less, and they produced about 25% of total U.S. milk volume.[18]

Annual premiums are calculated as the product of the premium rate per cwt. (from **Table 2**), the selected coverage percentage (25% to 90%), and the annual production history (APH).

MPP Premium= (Premium Rate) * (Coverage %) * (APH)

Thus, the premium rate varies with the size of the participating dairy operation (i.e., whether it has greater or less than 4 million lbs. of APH), the level of margin protection selected (from $4.00/cwt. to $8.00/cwt. in $0.50/cwt. increments), and the coverage level selected (25% to 90%).

Until final USDA regulations are available specifying the exact manner in which MPP premiums are to be calculated, there are two potential interpretations of how to calculate the 4-million-lb. APH threshold. One method would use the unadjusted APH for all calculations, while a second method would first adjust the APH for the selected coverage level before carrying out the premium calculations. These distinctions (described below) become increasingly important for lower coverage levels.

First Premium Calculation Method: Unadjusted APH

According to the first potential method, the APH is split into two components—the first 4 million lbs., and all marketings above 4 million lbs.—prior to considering the producer's selected coverage level. Dairy producers with a production history in excess of 4 million lbs. would be charged the lower premium rate on the first "unadjusted" 4 million lbs. and the higher premium rate on the amounts above that.

EXAMPLE OF MPP PREMIUM RATE CALCULATION USING THE UNADJUSTED APH

Suppose a dairy producer with an APH of 120,000 cwt. selects a \$6.50/cwt. MPP margin protection level with a 70% coverage level. The annual premium will be calculated as follows.

Unadjusted APH = 120,000 cwt.

For the first 4 million lbs. (or 40,000 cwt.) of APH, use the MPP premium rate of \$0.09/cwt.:

MPP Premium-1 = (\$0.09) * (70%) * (40,000 cwt.) = \$2,520

For all APH milk production above the first 4 million lbs. (or 120,000 cwt. – 40,000 cwt.) use the MPP premium rate of \$0.29/cwt.:

MPP Premium-2 = (\$0.29) * (70%) * (80,000 cwt.) = \$16,240

The **Total MPP Premium** is the sum: \$2,520 + \$16,240 = **\$18,760**.

This MPP premium is in addition to the annual administrative fee of \$100.

Second Premium Calculation Method: APH Adjusted for Coverage Level

According to the second potential method, the APH is first multiplied by the coverage level, before the adjustedAPH is split into two components—the first 4 million lbs., and the remaining adjustedAPH above 4 million lbs. This method restricts the premium calculations to APH that is actually covered by MPP. Dairy producers with an adjusted production history in excess of 4 million lbs. would be charged the lower premium rate on the first 4 million lbs. of adjusted APH and the higher premium rate on the amounts of adjustedAPH above that.

EXAMPLE OF MPP PREMIUM RATE CALCULATION USING THE ADJUSTED APH

Using the same example, suppose a dairy producer with an APH of 120,000 cwt. selects a $6.50/cwt. MPP margin protection level with a 70% coverage level. First, the APH is adjusted by the coverage level as:

Adjusted APH = APH * (coverage level) = (120,000 cwt.) * (70%) = **84,000 cwt**.

The annual premium will be calculated as follows.

For the first 4 million lbs. (or 40,000 cwt.) of APH, use the MPP premium rate of $0.09/cwt.:

MPP Premium-1 = ($0.09) * (40,000 cwt.) = $3,600.

For all adjusted APH milk production above the first 4 million lbs. (or 84,000 cwt. – 40,000 cwt.) use the MPP premium rate of $0.29/cwt.:

MPP Premium-2 = ($0.29) * (44,000 cwt.) = $12,760.

The **Total MPP Premium** is the sum: $3,600 + $12,760 = **$16,360.**

This MPP premium is in addition to the annual administrative fee of $100.

The differences in how to calculate the premium derive from the two interpretations of how to calculate the relevant portion of the APH as described in Section 1407 ("Premiums for Margin Protection Program") of the 2014 farm bill. The total premium owed under the second method represents a substantial reduction from the first method—$16,360 versus $18,760—and makes the choice of coverage level increasingly important in evaluating the cost as well as the potential benefits of the program.

Margin Coverage Distinction: $6.50/cwt. vs. $7.00/cwt.

Separately, it is worth noting that the premium structure strongly encourages participation at the $6.50/cwt. level for coverage greater than 4 million lbs., regardless of which method is used to calculate the premium. To raise the margin coverage level an additional $0.50/cwt. (from $6.50/cwt. to $7.00/cwt.), an additional $0.54/cwt. of premium must be paid (i.e., $0.83/cwt. -$0.29/cwt.). The total premium cost for insuring 90% of the APH of 120,000 cwt. at a margin of $7.00/cwt. for these same hypothetical scenarios would be $67,572 under the unadjusted APH method and $65,120 under the adjusted APH method—nearly quadruple the premium cost for $6.50/cwt. margin protection.

The timing and manner of premium payments is something that USDA will have to develop when it promulgates specific rules. The 2014 farm bill requires USDA to provide more than one method of payment and to use a method that "maximizes dairy operation payment flexibility and program integrity."

The 2014 farm bill also states that a participating dairy operation that fails to pay its administrative fee and/or premiums remains legally obligated for such charges, and may not receive any indemnities under MPP until such charges are fully paid.

THE DAIRY PRODUCT DONATION PROGRAM (DPDP)

According to the 2014 farm bill, no less than 120 days after the dairy margin protection program is operational, USDA shall establish and administer a Dairy Product Donation Program (DPDP) to (1) address low dairy operating margins, and (2) provide nutrition assistance to individuals in low-income groups.

A result from the dairy policy debate that preceded the final 2014 farm bill (see **Appendix**) was as follows. It is generally agreed that dairy operating margins are low when the equilibrium between milk supply and demand is out of balance with supply exceeding demand. The original proponents of the dairy margin program had argued that an additional policy feature—a supply disincentive component—should be added that would work to reduce supply and bring the milk supply-demand equation back into balance at a higher margin. Opponents of this concept argued that the government should not manage supplies, but should instead work to enhance demand to rebalance the milk supply and demand. This latter argument won out in conference when conferees proposed an entirely new "demand enhancing" program—DPDP.

DPDP Program Activities

USDA must announce and undertake DPDP activities whenever the actual dairy production margin (ADPM) is $4.00/cwt. or less for each of the immediately preceding 2 months. When DPDP has been triggered, USDA must immediately purchase dairy products, at prevailing market prices, until such time as one of the termination conditions (described below) is met.

USDA must consult with public and private nonprofit organizations—organized to feed low-income populations—in order to determine the types and quantities of dairy products to purchase under the DPDP. USDA is directed to use the funds, facilities, and authorities of the Commodity Credit Corporation (CCC) to implement and administer the DPDP.

DPDP Distribution of Purchases

The 2014 farm bill directs USDA to distribute, but not store, the dairy products purchased under DPDP so as to encourage greater domestic consumption by diverting them to persons in low-income groups as determined by USDA. USDA shall use public and private nonprofit organizations for distribution of DPDP dairy products to provide, without cost or waste, nutrition assistance to individuals in low-income groups. Any organization receiving dairy products under the DPDP may not sell the products back to commercial markets.

DPDP Program Termination Conditions

USDA must terminate its DPDP activities whenever any one of the following conditions occurs:

1. USDA has made DPDP purchases for three consecutive months, even if the **margin remains < $4.00/cwt.**;
2. the **margin is > $4.00/cwt.** for the preceding month;
3. when **$3.00/cwt. < margin < $4.00/cwt.** for the preceding month and either:

   a. the U.S. cheddar cheese price > 105% of world price, or
   b. the U.S. non-fat dry milk price > 105% of world price; or
4. the **margin < $3.00/cwt.** for the preceding month and either:
   a. the U.S. cheddar cheese price > 107% of world price, or b. the U.S. non-fat dry milk price > 107% of world price.

The 2014 farm bill gives USDA the latitude to determine which domestic and international prices of cheddar cheese and non-fat skim milk powder will be used to assess the DPDP termination conditions.

BUDGET OUTLAYS: HISTORICAL AND PROJECTED

USDA outlays for the major dairy support programs have trended downward since the 1980 farm bill period (**Table 3**). An outlook for strong dairy product prices for the next several years in the CBO May 2013 baseline accounts for the relatively small net outlay projections of $161 million over 5 years (FY2014-FY2018) and $250 million over 10 years (FY2014-FY2023) for the major dairy programs, assuming an extension of dairy policy as it existed under the 2008 farm bill.

Table 3. U.S. Dairy Programs,Historical and Projected USDA Outlays ($ millions)

Farm Bill	Fiscal Years	DPPSP	Market Loss Assistance	MILC	DEIP	Total
1980	FY1981 - FY1985	10,592	—	—	—	10,592
1985	FY1986 - FY1990	6,221	—	—	8	6,229
1990	FY1991 - FY1996	1,388	—	—	544	1,932
1996	FY1997 - FY2002	2,284	1,000	—	481	3,765
2002	FY2003 - FY2007	1,120	—	2,538	90	3,748
2008[a]	FY2008 - FY2012	280	290	1,091	28	1,688

Table 3. (Continued)

Farm Bill	Fiscal Years	DPPSP	Market Loss Assistance	MILC	DEIP	Total
CBO 5-year Projections for FY2014-FY2018						
CBO Baseline[b]	FY2014 - FY2018	27	11	99	25	161
P.L. 113-79	FY2014 - FY2018	—	—	—	—	+241
CBO 10-year Projections for FY2014-FY2023						
CBO Baseline[b]	FY2014 - FY2023	47	18	140	45	250
P.L. 113-79[c]	FY2014 - FY2023	—	—	—	—	+912

Sources: Historical data are assembled by CRS using various USDA data sources; projected data for FY2014-FY2023 are from the Congressional Budget Office (CBO), May 2013 Baseline for Farm Programs, May 14, 2013.

Notes: USDA's Commodity Credit Corporation total outlays do not include the implicit costs to consumers of tariff-rate quotas which limit access to cheaper international products. Also, there are no federal outlays for FMMOs other than for their administration.

a. Data for FY2012 are not final, while FY2013 data are not complete and have been excluded from this table.

b. Projections from the CBO May 2013 baseline, assuming continuation of current law.

c. CBO cost estimates of the effects on direct spending and revenues of the conference agreement on H.R. 2642, the Agricultural Act of 2014, as reported on January 27, 2014, as scored against CBO's May 2013 baseline. See CBO letter to Chairman Frank Lucas, Committee on Agriculture, U.S. House of Representatives, January 28, 2014.

According to CBO, replacing current dairy policy with the new dairy programs of the 2014 farm bill will result in projected additional budgetary outlays above baseline over the 5-year (FY2014-FY2018) and 10-year (FY2014-FY2023) periods of $241 million and $912 million, respectively, under the Agricultural Act of 2014 (P.L. 113-79).[19] In other words, the total 5-year and 10-year outlays projected by CBO under the new dairy programs are $402 million and $1.162 billion, respectively, thus continuing the downward trend in program outlays per farm bill period.[20]

Dairy Issues

Because the MPP and DPDP are new programs that differ notably from previous support programs, several questions have been raised concerning their implementation and potential cost, as well as their effect on U.S. and international dairy markets and consumers. History provides some indications but solid evidence will likely only emerge with implementation. Some of the more noteworthy related issues are briefly discussed here.

1. Will DPDP Activities Supplement or Offset Commercial Sales?

As Bozic et al. (2014) have noted:[21]

> The fundamental question is whether or not the end-users of donated products would have purchased them on commercial markets anyway. If these donations are strictly additive to total dairy usage then the amount of dairy products served in Food Banks or other settings is increased. If the donation displaces commercial purchases that would have been made with other cash resources, then total commercial sales of dairy products would actually decline.
>
> To the extent that these donations are going to programs that have limited resources and continuously unmet needs, it is not unreasonable to speculate that commercial displacement will be minimal.

2. Will MPP Work as a Safety Net? If So, with What Consequences?

Economic analysis indicates that, by focusing on the national average milk operating margin rather than strictly on the price of milk and offering a range of margin protection and coverage choices, U.S. dairy producers, regardless of their geographic location or management style, will be provided with substantial opportunity to manage their milk-production risk environment under the new dairy programs.[22]

Another related, but unanswered question is, if MPP is effective as a safety net, will its subsidized premiums crowd out private sector risk markets?

3. Will Production Incentives Result from MPP & DPDP?

The margin protection program will likely have a risk reduction effect, and the fixed premium structure likely has a significant federal subsidy component embedded in it—although the actual extent of implicit subsidy has yet to be assessed. This combination can be expected to provide some degree of production incentives to participating milk producers.

Overly strong production incentives can disrupt the traditional market signals that lead to reduced milk production via herd culling.[23] Furthermore, MPP provisions could inadvertently result in a policy framework that gives advantage to "lumpy" over "incremental" growth at the farm level as producers may choose to respond by starting new dairy operations at new locations to be able to enroll new milk production in the MPP.

The nature by which DPDP will be implemented and its subsequent effect on dairy markets is unknown at present. Will DPDP act as a catalyst for further milk production by artificially inflating the true underlying demand for dairy products, thus encouraging greater milk production and leading to lower prices and higher program payments down the road?

Unlike early versions of the Dairy SecurityAct—the precursor to the dairy subtitle contained in the 2014 farm bill—which contained a strong production disincentive component in the form of the proposed Supply Management Program, MPP has only a mild disincentive in the form of its exclusion of new milk production in excess of the national average milk production growth rate. If producers act in unison to expand production beyond normal annual growth rates, even that milk restriction is minimized.

4. Can the MPP Program be Gamed by Participants?

MPP does not follow sound insurance principles—premiums are not set to reflect the risk environment in milk and feed markets but instead are fixed over the life of the farm bill. This fixed premium structure combined with the ability of the producer to annually choose a margin insurance level creates incentives to adversely game the program to one's financial advantage.[24]

Publicly available information on futures contract prices from the Chicago Mercantile Exchange for milk and feed can be used to forecast near-term milk production margins that, in turn, indicate the probability of indemnity payments under various MPP coverage levels.

As a result, strong adverse gaming incentives exist for participants in the MPP. USDA has not yet announced the terms and conditions under which MPP signup will occur; however, economic analysis suggests that the adverse gaming incentives can be reduced or mitigated by specifying an earlier sign-up date for coverage decisions—for example, 60 to 90 days ahead of each year's program starting date of January 1.[25]

5. How Will MPP Payments be Distributed across Farm Sizes?

Economic analysis suggests that, under the fixed premium structure of the 2014 farm bill—with strong differentiation between premiums on the first 4 million lbs. and on marketings above 4 million lbs.—net expected benefits under MPP will closely align with those under the previous MILC program, but with slightly less skewed results. Farms with fewer than 100 cows would receive approximately 38% of the expected benefits (compared with 39% under MILC), while farms with over 1,000 cows would account for approximately 15% of benefits (compared with 9% under MILC) during low margin periods.

Furthermore, the same analysis concluded that, "only when higher coverage levels are deep in-the-money will payments per cwt. be more consistent across farm sizes and result in the distribution of net benefits following closely to the distribution of milk production."[26]

6. What Are the Trade Implications of MPP & DPDP?

The 2014 farm bill dairy provisions made several changes that have profound implications for U.S. and international trade agreements. First and foremost, by repealing the Dairy Product Price Support Program (DPPSP) and the Dairy Export Incentive Program (DEIP), U.S. dairy programs have made a significant step in better complying with its domestic support commitments under the World Trade Organization (WTO).

During the 17-year period from 1995 through 2011 (the most recent year for which notification data is available), the implicit program costs of DPPSP has represented the single largest U.S. domestic support outlay of any single program commodity accounting for $93.5 billion ($5.5 billion annually) or 39% of all U.S. amber box domestic support notifications to the WTO.[27]

Thus, the repeal of DPPSP is likely to have a substantial effect in lowering annual U.S. domestic support notifications to the WTO, and keeping U.S. domestic support outlays well below the United States' $19.1 billion ceiling.[28]

It is not yet clear how USDA will notify program outlays under MPP and DPDP—for example, possibly as amber box but exempt under the *de minimis* exclusion—but current CBO spending projections of about $1.2 billion over 10 years or about $120 million per year pale in comparison with the DPPSP's annual average notification of $5.5 billion. In addition, repeal of DEIP fulfills a general commitment by WTO members to end use of agricultural export subsidies, and brings greater pressure to bear on the European Union, Switzerland, Norway, and the rest of the world to act similarly.

Finally, U.S. dairy TRQs remain in place but are WTO legal as they were included in the original U.S. country schedule of tariffs and quotas that was approved and accepted by all WTO-member countries.

Given that regulations for the MPP program have yet to be published, and implementation specifics and producer participation remain unknown, it would be premature to speculate on any possible new trade effects in commercial markets. Furthermore, given the very favorable market conditions for the U.S. dairy sector with near record high milk prices in early 2014 coupled with prospects for relatively low feed costs, it is expected that market incentives will dominate producer decision making for the foreseeable future.

Conclusion

The answers to these questions may emerge only slowly. The new MPP and DPDP programs are likely to be implemented during a period of sustained strong milk prices, moderate to weak feed costs, and sustained large margins that, if realized, would minimize the impact of the new dairy programs **(Figure 3)**.[29]

Appendix. Debate behind the New MPP and DPDP

As mentioned earlier in the text ("Origins of the New Dairy Policy"), a version of the National Milk Producers Federation's (NMPF's) Feed the Future (FTF) proposal was introduced in the 112th Congress as H.R. 3062, The Dairy SecurityAct (DSA), by House Agriculture Committee Ranking Member

Collin Peterson on September 23, 2011. A modified version of DSA appeared as "Subtitle D—Dairy," in Title I of both the House-reported (H.R. 6083) and Senate-passed (S. 3240) farm bills of the 112th Congress.

Dairy Policy in the 113th Congress

In the 113th Congress, both the Senate and the House passed different farm bills (S. 954, passed by the Senate on June 10, 2013, and H.R. 2642, passed by the House on July 11, 2013) that proposed replacing the traditional set of dairy price and income support programs with a new margin-based income support program—called the Dairy Production Margin Protection Program (DPMPP) under S. 954, and the Dairy Producer Margin Insurance Program (DPMIP) under H.R. 2642.

The two proposed margin programs were similar in many respects; however, the Senate bill (but not the House bill) also included an accompanying market stabilization program—the Dairy Market Stabilization Program (DMSP), which, under certain conditions, would reduce payments to participating producers for their milk marketings when the margin falls below proposed statutory thresholds.[30]

Early versions of the House farm bill in the 113th Congress also included the DMSP as part of new dairy policy. However, during the House Agriculture Committee's markup of its first version of the 2013 farm bill (H.R. 1947) in May 2013, an amendment (H.Amdt. 228) was introduced by Representatives Goodlatte and Scott that proposed removing the DMSP from H.R. 1947 and making some minor adjustments to DPMPP. The Goodlatte-Scott amendment (GSA) was defeated in committee by a vote of 28 to 26. The amendment was reintroduced during the House floor debate of H.R. 1947 and passed by a vote of 291-135 (May 15, 2013).

However, the full House voted to reject the amended farm bill (195-234) on June 20, 2013. On July 11, 2013, the full House passed (by a vote of 216-208) a second version of the 2013 farm bill (H.R. 2642) which included the Goodlatte-Scott amendment—thus removing DMSP and replacing DPMPP with DPMIP.

Conference Committee Resolves Differences

The differences between the Senate- and House-passed farm bills were resolved when conferees reported a conference agreement on January 27, 2014 (the Agricultural Act of 2014, H.R. 2642/H. Rept. 113-333). The dairy supply management debate was resolved when the conferees included an entirely new program—the Dairy Product Distribution Program (DPDP)—focused on demand enhancement, rather than supply control. Prior to the conference, the dairy debate had been limited to whether a supply management component should be included in the final bill— the concept of a DPDP did not emerge until conference. The full House and Senate approved the conference agreement on January 29 and February 4, respectively, and the President signed the measure into law on February 7, 2014 (P.L. 113-79).

Table A-1. Title I, Subtitle D- Dairy:Comparison of 2014 Farm Bill with Prior Law

Prior Law/Policy—Dairy	Enacted 2014 Farm Bill (P.L. 113-79)
Dairy Programs	
Repeal or Reauthorization of Dairy Programs	
Dairy Product Price Support Program. Mandates the direct support of cheese, nonfat dry milk, and butter at specified prices for five years (through December 31, 2012). Specifies minimum purchase prices of: block cheese, $1.13/lb.; barrel cheese, $1.10/lb.; butter, $1.05/lb.; and nonfat dry milk, $0.80/lb (same levels previously used to support the farm price of milk at $9.90 per hundred lbs. or hundredweight (cwt.)) Allows USDA sale of acquired products when market prices rise to 110% of purchase price. Allows reduction of mandated purchase prices when USDA acquisitions exceed specified levels. Expired on December 31, 2013, following a one-year extension. ***[7 U.S.C. 8771]***	Repealed. ***[Sec. 1421]***
Milk Income Loss Contract (MILC) Program. MILC is a counter-cyclical payment program. When the monthly farm price of fluid milk falls below	**Milk Income Loss Contract (MILC) Program**. Although extended temporarily, MILC is repealed effective the earlier of: the date the

Prior Law/ Policy—Dairy	**Enacted 2014 Farm Bill (P.L. 113-79)**
$16.94/cwt., all dairy farmers are paid an amount equal to 45% of the difference between $16.94 and the lower market price. Payments per farm are limited to the first 2.985 million lbs. of annual production. For the month of September 2013, the payment factor and the payment quantity were 34% and 2.4 million lbs., respectively. The $16.94/cwt. threshold price must be adjusted upward whenever feed costs are above $7.35/cwt. Beginning on September 1, 2013, the Nat'l. Avg. Dairy Feed Ration Cost trigger rises from $7.35/cwt. to $9.50/cwt. MILC program expired September 30, 2013. ***[7 U.S.C. 8773]***	new Margin Protection Program (MPP, see below) is operational or September 1, 2014. During the MILC extension period the payment rate shall be calculated using the pre-September 2013 parameters of 2.985 million lbs. for the payment limit, a $7.35/cwt. feed ration threshold, and a 45% rate as applied to both (1) the difference between the $16.94/cwt. threshold price (as adjusted by feed ration cost estimates) and the lower market price, and (2) the extent to which the national average dairy feed ration cost exceeds $7.35/cwt. to be used to adjust upward the $16.94/cwt. threshold price. ***[Sec. 1422]***
Dairy Export Incentive Program. Provides cash bonus payments to U.S. dairy exporters, subject to World Trade Organization obligations to limit export subsidies. Intended to counter foreign (mostly EU) dairy subsidies. Expired September 30, 2013. ***[15 U.S.C. 713a-14]***	Repealed. ***[Sec. 1423]***
Dairy Forward Pricing Program. Authorizes a dairy forward pricing program. Prices paid by milk handlers under the contracts are deemed to satisfy the minimum price requirements of federal milk marketing orders. Applies only to milk purchased for manufactured products (Classes II, III, and IV), and excludes milk purchased for fluid consumption (Class I). Expired on September 30, 2013. ***[7 U.S.C. 8772]***	Extended through FY2018. Allows for new contracts until September 30, 2018, but no contract can extend beyond September 30, 2021. ***[Sec. 1424]***
Dairy Indemnity Program. Authorizes payments to dairy farmers when a public regulatory agency directs removal of their raw milk from the market because of contamination by pesticides, nuclear radiation or fallout, or toxic substances and other chemical residues. Expired December 31, 2013. ***[7 U.S.C. 4501]***	Extended through FY2018. ***[Sec. 1425]***

Table A-1. (Continued)

Prior Law/ Policy—Dairy	Enacted 2014 Farm Bill (P.L. 113-79)
Dairy Promotion and Research Program. The Dairy Producer Stabilization Act of 1983 authorized a generic dairy product promotion, research, and nutrition education program, funded by a mandatory $0.15/cwt assessment on milk produced/marketed in the 48 contiguous states. Importers in all 50 states, the District of Columbia, and Puerto Rico must also pay an assessment rate of $0.075/cwt. on the milk equivalent in imported dairy products. Authorizes USDA to issue regulations on time and method of importer payments. Expired September 30, 2013. ***[7 U.S.C. 4504]***	Extended through FY2018. ***[Sec. 1426]***
Inclusion of California as Separate Federal Marketing Order. Upon the petition and approval of California dairy producers in the manner provided in section 608c of this title, the Secretary shall designate the state of California as a separate Federal milk marketing order (FMMO). The order covering California shall have the right to reblend and distribute order receipts to recognize quota value. USDA was required to announce any such proposed amendment to FMMOs by April 4, 1998, and to implement such amendment by April 4, 1999. ***[7 U.S.C. 7253(a)(2)]***	**Inclusion of Additional Order.** Adds a provision to remove the expiration date for USDA to act upon a petition from California dairy producers for designation as a separate FMMO. ***[Sec. 1410(d)]***
Federal Milk Marketing Order Review Commission. As established by the 2008 farm bill ***[Sec. 1509]***, the FMMO Review Commission is mandated to conduct a comprehensive review and evaluation of (1) FMMO system, and (2) non-FMMO systems.	Repealed. ***[Sec. 1427]***
Definitions	
No comparable provision.	**Actual Dairy Production Margin:** difference between all-milk price and average feed cost. ***[Sec. 1401(1)]***

Prior Law/ Policy—Dairy	**Enacted 2014 Farm Bill (P.L. 113-79)**
No comparable provision.	**All-Milk Price:** the national average price received, per cwt. of milk, by dairy operations. ***[Sec. 1401(2)]***
No comparable provision.	**Average Feed Cost:** the average price paid for feed used by a dairy operation to produce a cwt. of milk, as determined by the formula—1.0728 x (corn price per bu.) + 0.00735 x (soybean meal price per ton) + 0.0137 x (alfalfa hay price per ton). ***[Sec. 1401(3)]***
No comparable provision.	**Consecutive 2-Month Period:** the six 2-month periods of Jan.-Feb., Mar.-Apr., May-June, July-Aug., Sep.-Oct., and Nov.-Dec. ***[Sec. 1401(4)]***
No comparable provision.	**Dairy Operation.** An individual or entity that shares in the risk of producing milk and contributes land, labor, management, equipment, or capital to the dairy operations. ***[1401(5)]***
No comparable provision.	**Calculation of Average Feed Costs:** Corn and alfalfa hay prices are monthly national average prices received as reported by USDA in *Agricultural Prices.* The soybean meal price is the monthly price for central Illinois as reported by USDA in *Market News*. ***[Sec. 1402(a)]***
No comparable provision.	**Calculation of Actual Dairy Production Margin**: The margin is calculated for each 2-month period as the difference between the 2-month average all-milk price and the 2-month average feed cost. ***[Sec. 1402b(1)],*** to be made as soon as practicable using the full-month price of the applicable reference month. ***[Sec. 1402(b)(2)]***
Margin Protection Program (MPP) for Dairy Producers	
No comparable provision.	**Establishment of Margin Protection Program (MPP) for Dairy Producers.** USDA shall establish a margin protection program for dairy producers no later than September 1, 2014. ***[Sec. 1403]***

Table A-1. (Continued)

Prior Law/ Policy—Dairy	Enacted 2014 Farm Bill (P.L. 113-79)
No comparable provision.	**Duration:** The margin protection program ends on December 31, 2018. ***[Sec. 1409]***
No comparable provision.	**Eligibility for MPP.** All dairy producers are eligible to participate. ***[Sec. 1404(a)]***
No comparable provision.	**Registration:** USDA shall specify the manner and form by which a participating dairy operation may register to participate in the margin program. ***[Sec. 1404(b)(1)]***
No comparable provision.	**Treatment of Multi-Producer Dairy Operations.** If a dairy operation consists of more than one dairy producer, all of the dairy producers shall be treated as a single dairy producer for purposes of participation in the margin protection program. ***[Sec. 1404(b)(2)]***
No comparable provision.	**Treatment of Producers with Multiple Dairy Operations.** If a dairy producer operates 2 or more dairy operations, each dairy operation shall separately register to participate in the margin protection program. ***[Sec. 1404(b)(3)]***
No comparable provision.	**Annual administration fee.** An annual administration fee of $100 is required per dairy operation participating in MPP. ***[Sec. 1404(c)(2)]***
No comparable provision. **LGM-D** is an insurance program, operated by USDA's Risk Management Agency, under the permanent authority of the Federal Crop Insurance Act, as amended (7 U.S.C. 1501 et seq.).	**Relationship to Livestock Gross Margin for Dairy Program (LGM-D):** A dairy operation may participate in either MPP or the LGM-D, but not both. ***[Sec. 1404(d)]***
No comparable provision.	**Production History.** At the initial registration, the production history is equal to the highest annual milk marketings of the dairy operation during any one of the 3 calendar years

Prior Law/ Policy—Dairy	**Enacted 2014 Farm Bill (P.L. 113-79)**
	2011, 2012, or 2013. In subsequent years, USDA shall adjust the production history to reflect any increase in the national average milk production. ***[Sec. 1405(a)]*** Special provisions are made for new dairy operations. ***[Sec. 1405(b)]***
No comparable provision.	**MPP Coverage Level Threshold and Coverage Percentage.** Dairy producers are offered the annual choice of purchasing margin insurance with coverage levels ranging in $0.50/cwt. increments from a minimum of $4.00/cwt. to a maximum of $8.00/cwt.
	A participating producer shall elect a coverage percentage available in 5% increments equal to not more than 90%, nor less than 25% of the **Production History** of the dairy operation. ***[Sec. 1406(a)]***
No comparable provision.	**MPP Payment Threshold.** A payment is made to participating dairy operations whenever the 2-month average actual dairy production margin is less than the coverage level threshold selected by the producer. ***[Sec. 1406(b)]***
	MPP Payment Rate. The MPP rate equals the amount that the margin is below the selected margin coverage level threshold. The **MPP Total Payment** equals the payment rate x the selected coverage % x (APH/6). ***[Sec. 1406(c)]***
No comparable provision.	**Producer Premiums for MPP.** An annual premium must be paid, equal to the product of the selected coverage %, the production history, and the premium rate per cwt. of milk. ***[Sec. 1407(a)]***
	The **MPP Premium Rate** schedule varies based on the scale of operations and the selected coverage %.

Table A-1. (Continued)

Prior Law/ Policy—Dairy	Enacted 2014 Farm Bill (P.L. 113-79)
	For the first 4 million lbs. of milk marketings the premium per cwt. is $0.00 for $4.00 margin coverage; $0.01 for $4.50; $0.025 for $5.00; $0.040 for $5.50; $0.055 for $6.00; $0.09 for $6.50; $0.217 for $7.00; $0.30 for $7.50; and $0.475 for $8.00. For calendar 2014 and 2015, all of the preceding premiums (except the $8.00/cwt. coverage level) are reduced by 25%. ***[Sec. 1407(b)]***
	In excess of 4 million lbs. the premium per cwt. is: $0.00 for $4.00; $0.020 for $4.50; $0.040 for $5.00; $0.10 for $5.50; $0.155 for $6.00; $0.29 for $6.50; $0.83 for $7.00; $1.06 for $7.50; and $1.36 for $8.00. ***[Sec. 1407(c)]***
No comparable provision.	**Time for Premium Payments.** USDA is instructed to provide more than one method of payment and to use a method that "maximizes dairy operation payment flexibility and program integrity." ***[Sec. 1407(d)]***
No comparable provision.	The premium is pro-rated for new dairy producers and may be waived in the case of death, retirement, permanent dissolution, or other circumstances as judged by USDA. ***[Sec. 1407(c)]***
No comparable provision.	**Failure to Pay Administrative Fees or Premiums.** A producer that fails to pay the administrative fee or premium remains legally obligated to pay and may not receive a MPP payment until all obligations are paid. USDA may take such action as necessary to collect any unpaid obligations. ***[Sec. 1408]***
Rulemaking Requirements. Standard rulemaking procedures generally require federal agencies to issue notices and take	**Administration and Enforcement.** USDA is required to promulgate regulations to address administrative

Prior Law/ Policy—Dairy	Enacted 2014 Farm Bill (P.L. 113-79)
comments on proposed rules. ***[5 U.S.C. 553(b)]***	and enforcement issues in carrying out the margin protection program. ***[Sec. 1410(a)]***
	Reconstitution of Dairy Operation. USDA is required to promulgate regulations to address administrative and enforcement issues and prohibit reconstitution of a dairy operation for the purpose of the dairy producer receiving margin protection payments. ***[Sec. 1410(b)]***
	Administrative Appeals. USDA is required to promulgate regulations to provide for administrative appeals of USDA decisions that are adverse to participants of the margin protection program. ***[Sec. 1410(c)]***
Dairy Product Donation Program (DPDP)	
No comparable provision.	**Establishment of a Dairy Product Donation Program (DPDP).** No less than 120 days after the dairy margin protection program is operational, USDA shall establish and administer a DPDP to: (1) address low dairy operating margins, and (2) provide nutrition assistance to individuals in low-income groups. ***[Sec. 1431(a)]***
No comparable provision.	**DPDP Program Trigger.** USDA shall announce and undertake DPDP activities whenever the actual dairy production margin is $4.00/cwt. or less for each of the immediately preceding 2 months. ***[Sec. 1431(b)]***
No comparable provision.	**DPDP Program Activities.** When DPDP has been triggered, USDA shall immediately purchase dairy products, at prevailing market prices, until such time as one of the termination conditions is met.
	USDA shall consult with public and private nonprofit organizations organized to feed low-income populations in order to determine the types and quantities of dairy products

Table A-1. (Continued)

Prior Law/ Policy—Dairy	Enacted 2014 Farm Bill (P.L. 113-79)
	to purchase under the DPDP. ***[Sec. 1431(c)]***
	The funds, facilities, and authorities of the Commodity Credit Corporation shall be available to USDA to implement and administer the DPDP. ***[Sec. 1431(g)]***
No comparable provision.	**DPDP Program Termination.** USDA shall cease DPDP activities whenever any one of the following conditions occurs: (1) USDA has made DPDP purchases for three consecutive months, even if the margin remains < $4.00/cwt.; (2) the margin > $4.00/cwt. for the preceding month; (3) when $3.00/cwt. < margin < $4.00/cwt. for the preceding month and either: (a) the U.S. cheddar cheese price > 105% of world price, or (b) the U.S. non-fat dry milk price > 105% of world price; or (4) the margin < $3.00/cwt. for the preceding month and either: (a) the U.S. cheddar cheese price > 107% of world price, or (b) the U.S. non-fat dry milk price > 107% of world price. For this section, USDA shall determine the domestic and international prices of cheddar cheese and non-fat skim milk powder. ***[Sec. 1431(d)]***
No comparable provision.	**Distribution of DPDP Purchases.** USDA shall distribute, but not store, the dairy products purchased under DPDP so as to encourage domestic consumption by diverting them to persons in low-income groups as determined by USDA. USDA shall use public and private nonprofit organizations for distribution of DPDP dairy products to provide, without cost

Prior Law/ Policy—Dairy	Enacted 2014 Farm Bill (P.L. 113-79)
	or waste, nutrition assistance to individuals in low-income groups. ***[Sec. 1431(e)]*** Any organization receiving dairy products under the DPDP may not sell the products back to commercial markets. ***[Sec. 1431(f)]***
No comparable provision.	**DPDP Duration.** The DPDP shall end on December 31, 2018. ***[Sec. 1431(h)]***

Source: Title I- Commodities; Subtitle D—Dairy, The Agricultural Act of 2014; P.L. 113-79.

Notes: For a comparison of prior and enacted law with the provisions in the House and Senate versions of the 2014 farm bill (i.e., the Senate-Passed S. 954 and the House-Passed H.R. 2642) see CRS Report R43076, *The 2014 Farm Bill (P.L. 113-79): Summary and Side-by-Side.*

End Notes

[1] This report borrows, in part, from the discussion of new dairy programs in "The Dairy Subtitle of the Agricultural Act of 2014," PDMP Information Letter 14-01, by Marin Bozic, John Newton, Andrew M. Novakovic, Mark Stephenson, and Cameron S. Thraen, January 31, 2014.

[2] For a comparison of new 2014 farm bill dairy policy provisions with previous law as well as with both the House and Senate farm bill proposals—the Senate-passed S. 954 and the House-passed H.R. 2642—see CRS Report R43076, *The 2014 Farm Bill (P.L. 113-79): Summary and Side-by-Side*.

[3] The U.S. government purchased storable dairy commodities in 1933 and 1941 as a way to shore up farm milk prices and provide food for needy families. During WWII, the same mechanism was used to ensure adequate production. In the tumultuous economic aftermath of WWII, this means of supporting farm milk prices was made permanent through the Dairy Price Support Program of the Agricultural Act of 1949.

[4] The original program—named the Dairy Price Support Program—had a statutorily determined support price for fluid milk (e.g., $9.90 per cwt. in the mid-2000s). The program was renamed by the 2008 farm bill when direct fluid milk price support was shifted to indirect support via government purchases of manufactured products including butter, cheese, and milk powder at statutorily established prices.

[5] USDA, Economic Research Service (ERS), Commodity Costs and Returns data, retrieved on July 23, 2012, from http://www.ers.usda.gov/data-products/milk-cost-of-production-estimates. aspx.

[6] See the NMPF Foundation for the Future website at http://www.futurefordairy.com/.

[7] More detailed descriptions of the previous-law dairy programs, DPPSP, MILC, and DEIP, are available in CRS Report R42736, *Dairy Policy Proposals in the Next Farm Bill* and CRS Report RL34036, *Dairy Policy and the 2008 Farm Bill.*

[8] See USDA, AMS, Dairy Programs, "Federal Milk Marketing Orders," listed under "Programs and Services" at http://www.ams.usda.gov/AMSv1.0/dairy.

[9] For details by product, see the Harmonized Tariff Schedule of the United States (2012) (rev. 2), Chapter 4, pp. 2-7.

[10] For a detailed discussion of "permanent farm law" see CRS Report R42442, *Expiration and Extension of the 2008 Farm Bill.*

[11] For more information on the LGM-D insurance program, see http://www.rma.usda.gov/pubs/rme/lgmdairy.pdf.

[12] USDA is required to publish a notice in the *Federal Register*, to inform dairy producers and other stakeholders of the availability of the new MPP and DPDP programs.

[13] Monthly prices received by farmers for their milk, and the cost of their primary feedstuffs, corn, and alfalfa hay, are published monthly in *Agricultural Prices*, National Agricultural Statistics Service (NASS), USDA. The average wholesale price for soybean meal, another essential component of the dairy feed ration, Central Illinois, is reported in *Market News,* Agricultural Market Service (AMS), USDA.

[14] The feed cost formula is intended to represent the dairy ration that would be consistent with recommended nutrition to pounds of milk including the dairy cow and the herd complement of dry cows, hospital cows and youngstock at the average U.S. milk yield per cow. For a detailed description of the feed cost formula derivation, see *Foundation for the Future*, NMPF, June 2010, pp. 16-19; at http://www.futurefordairy.com/.

[15] This differs from the LGM-D program mentioned earlier. LGM-D offers similar margin insurance, but it allows the producer to select different quantities of two feeds; corn and soybean meal. In addition, LGM-D uses the Federal Order Class III futures price reported on the Chicago Mercantile Exchange, and not the USDA-NASS reported U.S. All-Milk price.

[16] USDA is instructed to use the fees collected under MPP to cover administrative costs incurred in carrying out the program.

[17] For more information on the LGM-D insurance program, see http://www.rma.usda.gov/pubs/rme/lgmdairy.pdf.

[18] Farm-size shares are from "Farms, Land in Farms, and Livestock Operations," NASS, USDA, February 17, 2012, and total milk production is from "Milk Production, Disposition, and Income," NASS, USDA, April 25, 2012.

[19] See CBO letter to Chairman Frank Lucas, Committee on Agriculture, U.S. House of Representatives, January 28, 2014.

[20] These estimates are the sum of the baseline score (which assumed a continuation of the 2008 farm bill) and the additional cost attributable to the new dairy programs 2014 farm bill.

[21] M. Bozic et al. "The Dairy Subtitle of the Agricultural Act of 2014," PDMP Information Letter 14-01, Jan. 31, 2014.

[22] John Newton and Cam Thraen, "The Dairy Safety Net Debate of 2013 Part III: The Compromise Dairy Safety Net Solution," *farmdocdaily.com*, Dept. of Agr. and Cons. Econ., Univ. of Illinois Urbana-Champaign, Feb. 12, 2014.

[23] M. Bozic et al. "The Dairy Subtitle of the Agricultural Act of 2014," PDMP Information Letter 14-01, Jan. 31, 2014.

[24] Ibid.

[25] Ibid.

[26] Ibid., p. 3.

[27] The dairy price support program notifications have been so large primarily because they have been measured by a formula that equals the difference between domestic support price of \$9.90/cwt. and a world market price of \$7.25/cwt. (which was fixed at the 1994 value as part of the original WTO agreement) times annual U.S. milk production; but not by how much in federal budgetary outlays were spent to support domestic milk prices.

[28] For more information, see CRS Report RS20840, *Agriculture in the WTO: Limits on Domestic Support*

[29] For recent long-term dairy market projections, see Food and Agricultural Policy Research Institute (FAPRI), "U.S. Baseline Briefing Book", FAPRI-MU Report #02-14, March 13, 2014; at http://www.fapri.missouri.edu/.

[30] For a detailed discussion of differences in dairy provisions between the House-passed H.R. 2642 and the Senate-passed S. 954, see CRS Report R42736, *Dairy Policy Proposals in the Next Farm Bill.*

In: U.S. Dairy Support Programs
Editor: Carla Edie

ISBN: 978-1-63321-715-7

Chapter 2

LIVESTOCK GROSS MARGIN-DAIRY INSURANCE: AN ASSESSMENT OF RISK MANAGEMENT AND POTENTIAL SUPPLY IMPACTS*

Kenny Burdine, Roberto Mosheim, Don P. Blayney and Leigh J. Maynard

ABSTRACT

Public risk management policies for dairy producers have the potential to induce expansion in milk supplies, which might lower farm-level prices and offset risk-reduction benefits. An evaluation of USDA's Livestock Gross Margin-Dairy (LGM-Dairy) insurance program finds economic downside risk significantly reduced, with potential to induce modest supply expansion (0 to 3 percent) if widely adopted. Supply impacts are likely limited due to relatively low participation levels and a minimal ("inelastic") supply response to risk. LGM-Dairy is more flexible and convenient than other risk management tools, such as hedging directly in futures or options markets, especially for small farms.

* This is an edited, reformatted and augmented version of Economic Research Report Number 163, issued by the USDA Economic Research Service, March 2014.

Keywords: Dairy, gross margins, risk management, LGM-Dairy, insurance, milk supplies, livestock

What Is the Issue?

Livestock Gross Margin-Dairy (LGM-Dairy) is a risk management tool that enables dairy producers to purchase insurance against decreases in gross margins—the difference between revenues and feed costs. LGM-Dairy offers features seen in only a few other insurance and risk management tools. As a type of index insurance that makes payments based on broader market conditions rather than individual farm outcomes, it reduces the likelihood of common insurance problems such as riskier behavior by the insured to increase indemnities ("moral hazard") or information gaps that make it difficult for the insurer to determine appropriate premiums ("adverse selection"). It also gives dairy producers the ability to insure a margin, not just the output price. Lastly, LGM-Dairy is available through local insurance agents and is offered in scales appropriate for any size dairy operation. Less appealing to producers may be the very short buying period for an LGM-Dairy contract and the program's rules and specifications.

The introduction of LGM-Dairy in 2008 coincided with increased market volatility in both input (feed) and output (dairy) prices, making it an attractive risk management tool. However, funding allocated to all Federal Crop Insurance Corporation livestock insurance programs for expenses that include administrative and operating (A&O) subsidies and premium subsidies paid by the Government on behalf of participating producers is limited to $20 million per reinsurance year, which prevents producers from using LGM-Dairy insurance as a year-round risk management strategy.

Initial and ongoing 2012 farm bill legislative efforts included a dairy margin insurance plan similar to LGM-Dairy but with different operating parameters. Among the key differences are the levels of margins to be insured and the determination of premium rates. Information on the effectiveness of the existing LGM-Dairy program as a risk management tool and its impact on milk supplies will help inform continued efforts to understand the impacts of policies for dairy risk management.

What Did the Study Find?

Among the key findings of the economic analysis are:

- Had LGM-Dairy been widely available during the entire period studied (2001-11), reductions in risk would have ranged from 28 to 39 percent in the 13 dairy-producing regions analyzed.
- While the risk reductions were significant, use of LGM-Dairy was shown to have little effect on the average producer margins across regions, which would have changed by -1 to 2 percent.
- Based on the estimated risk-reduction levels, and risk elasticities derived from the literature and risk modeling, the potential for LGM-Dairy to induce an expansion in milk supplies, if it had been more widely available and adopted, was modest across regions, ranging from 0 to 3 percent. The upper end of this range is unlikely because producer participation has been far below 100 percent to date, and because the upper estimate requires producer supplies to respond strongly to changes in risk. The main incentive for producers to expand supply comes from subsidies currently offered on the LGM-Dairy insurance premiums.

Generally, results suggest that insuring both output prices and feed costs is more effective for managing risk than insuring output prices alone and appears to produce similar risk management performance across the selected regions. LGM-Dairy, as a gross margin insurance program, is more flexible in terms of coverage than other risk management tools, such as hedging in futures or options markets, which likely makes LGM-Dairy more attractive for small farms.

How Was the Study Conducted?

The study consisted of two parts, the first of which was an analysis of the effectiveness of LGM-Dairy as a risk management tool. Using monthly cash and futures prices for milk, corn, and soybean meal, researchers calculated gross margins (milk price minus feed costs) with and without the use of LGM-Dairy as a risk management tool. Although the program was established in 2008, outcomes were simulated using data from January 2001 to April 2011.

The second component was estimating the likely effects of risk reduction on milk production. Researchers collected milk production data from multiple sources for a subset of the period used above (first quarter 2006 to fourth quarter 2010). Regression analysis was employed to estimate production response to market risk. Results from supply modeling were used in conjunction with results from risk estimation to evaluate the likely impacts on production.

INTRODUCTION

Over the last 20 years, producers of all agricultural products have increasingly faced challenges stemming from price volatility. Dairy producers have faced rapidly changing milk prices and input prices, primarily for feeds, and the associated impacts of those changes on profitability. The monthly average U.S. all-milk price was highly volatile from 1990 to 2012, particularly during the later years of that period (figure 1). Factors that may account for the increasing variability in milk prices include increased U.S. involvement in, and dependence on, export markets and weather events in both the United States and Oceania (Australia and New Zealand) that affected production and dairy stock levels. More recently, dairy producers also faced higher feed costs that affected operational and investment decisions.

Traditional price risk management tools, such as forward contracting and the use of futures and options markets, present opportunities to manage the risks associated with price volatility, but they also present challenges. Dairy producers generally have struggled to adopt futures and options trading as a means of price protection. The futures contracts available are only a cross-hedge opportunity for the all-milk price, with a very uncertain basis risk (see glossary). And, scale issues often prevent smaller dairy operations from using milk futures and options (Harwood et al., 1999). The Livestock Gross Margin for Dairy (LGM-Dairy) insurance program was initially developed in response to these challenges.

LGM-Dairy is a relatively new public risk management program overseen by USDA's Risk Management Agency (RMA). It was designed to reduce the negative effects of milk and feed price volatility on U.S. dairy farms (USDA, Risk Management Agency, 2010a) and was first made available in June 2008 for the 2009 livestock reinsurance year and has been available for each reinsurance year since that time.

This study explores whether LGM-Dairy is likely to induce an expansion in the supply of milk if widely adopted. The aggregate demand for farm-level milk is price-inelastic, implying that aggregate dairy farm revenues fall when the milk supply increases (i.e., the supply curve shifts out (to the right)). The study's authors compare historical gross margins over several years with the corresponding outcomes had LGM-Dairy been employed to estimate the degree of risk reduction. They also use supply impact parameters from the literature (Chavas and Holt, 1990; Bakhshi and Kerr, 2009; Lin and Dismukes, 2007) and from original supply modeling within this work to estimate the impact on milk supplies corresponding to the estimated risk reduction level under LGM-Dairy.

BACKGROUND

Until recently, output (milk) price risk management was not a key element of decisionmaking by milk producers. This stems from public policies initiated in the 1930s that have generally precluded the need for dairy farmers to address issues related to risk management. Until the passage of the Agricultural Act of 2014, those policies and programs remained in place today, although in modified forms.[1] One such program, Dairy Price Support (DPS), was established in the 1949 Farm Act and lasted in its original form until enactment of the 2008 Farm Act.

The DPS program essentially provided downside price protection by supporting producer milk prices at or above a specified level. This price support was achieved through Government purchases of "surplus" dairy products (American cheese, butter, and nonfat dry milk) at prices announced by USDA, effectively removing the products from the domestic market. While not necessarily preventing milk prices from falling below the support price, the program did help ensure that such low prices would not persist. The program truncated the milk price distribution at the support price level, at least until the early 1990s. At that time, the support prices began to be phased down to such low levels that they rarely trigger sales to the Government and thus provide minimal downside risk protection.

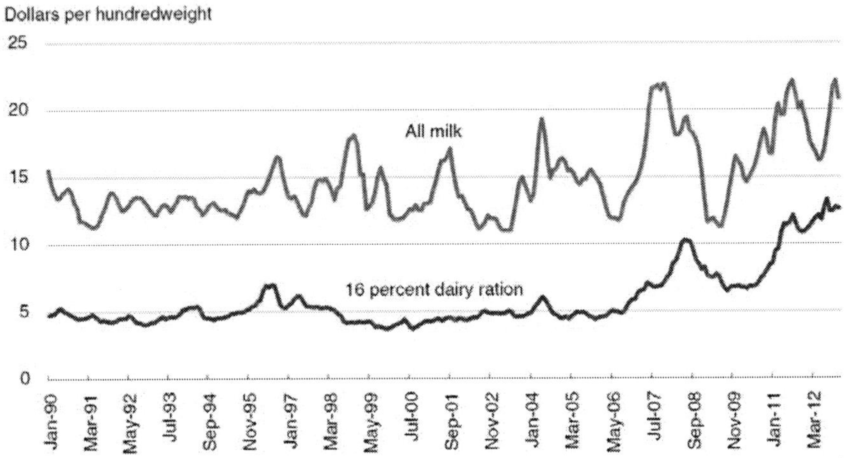

Sources: USDA, Economic Research Service using data from USDA, National Agricultural Statistics Service and Gould (2011).

Figure 1. Average U.S. all-milk price and feed costs.

Other policies and programs that have been implemented were designed to provide dairy producers with added income if milk prices were low. These programs included dairy herd buyouts, the Milk Income Loss Contract (MILC), and the Dairy Export Incentive Program (DEIP).

The Dairy Options Pilot Program

Over time, agricultural price policy prescriptions have shifted from direct payments and price supports toward promoting the use of market-based price risk management tools (FAPRI, 2012). Futures and options strategies are commonly used in some agricultural sectors; however, futures and options markets are generally thin for dairy products. In 1999, a Dairy Option Pilot Program (DOPP) was launched to educate milk producers about the use of options and thereby increase the effectiveness of these risk management tools. Findings suggest that DOPP did increase the volume of dairy options traded but also increased their price (Bushena and McNew, 2005).

Other studies suggest that risk reduction could be achieved through the use of dairy futures and options but that several factors limit the effectiveness of such a strategy. First, hedging is most effective in those areas where milk use in cheese production, referred to as class III milk, was greatest; contracts for class IV milk, the milk used for butter and nonfat dry milk production, are generally more thinly traded. Second, findings on futures market indicators, particularly hedge ratios (Stoll and Whaley, 1993), by Maynard et al. (2005) suggest that use of futures and options as a risk management strategy is most practical for very large (high volume) operations. Third, there continues to be a perception that existing public milk pricing policy limits the need for price risk management (Maynard et al., 2005).

Lastly, how does one assess the degree to which the DOPP met its goal of educating dairy producers. Such an assessment is always a challenge when introducing a new marketing tool. Previous research suggests that comprehensive training on put options increased the comfort level of producers, but many producers felt they required even more training. Also, many producers still saw using futures and options as a gamble rather than a price risk management tool (Ibendahl et al., 2002).

Livestock Gross Margin-Dairy Insurance Program

The LGM-Dairy program is an insurance product that provides compensation to dairy producers for qualifying losses paid under the product's specifications (for more information on the program, see box, "How Does LGM-Dairy Work?"). It is one of the array of price risk management tools for livestock provided under the auspices of the Federal Crop Insurance Act, as amended, and offers features seen in few other previous insurance and risk management products. First, LGM-Dairy is a type of index insurance, a characteristic that reduces the likelihood of two common problems associated with insurance—moral hazard and adverse selection (Barnett, 2004). In the case of LGM-Dairy, indemnities are based on futures prices for class III milk, corn, and soybean meal rather than on actual prices received for milk and paid for the feed inputs.[2]

Second, unlike futures and options on milk prices alone, LGM-Dairy offers dairy producers the opportunity to insure a margin similar to employing a "bundled option strategy" (see glossary). The milk price is only one piece of a milk producer's measure of profitability and while basis risk certainly remains, the mechanics of LGM-Dairy provide protection from both a decrease in milk prices and an increase in feed costs.

Third, LGM-Dairy availability through insurance agents rather than commodity futures brokers is perhaps a more preferable arrangement for dairy producers, and the product is offered in levels of coverage appropriate for any size dairy. Finally, increased dairy market volatility preceded the introduction of LGM-Dairy, which made the product more attractive as a risk management tool.

However, the capacity of the LGM-Dairy insurance program is limited. Funds allocated to the livestock insurance programs offered by the Federal Crop Insurance Corporation (FCIC), including LGM-Dairy, have been legislatively capped at $20 million per reinsurance (fiscal) year for administrative and operating (A&O) subsidies, an amount that could potentially be depleted quite quickly. In fact, the FCIC funds initially allocated to livestock insurance programs for the 2012 reinsurance year were exhausted by December 2011, with sales not expected to resume until October 2012 (USDA, Risk Management Agency, 2011). The various RMA livestock insurance programs, including LGM-Dairy, were reauthorized for the 2013 reinsurance year. It has been suggested that the limited funding and inconsistent availability have prevented producers from using LGM-Dairy insurance on a regular basis (Wright, 2012).

LGM-Dairy was first made available for the 2008 reinsurance year (October 2007-September 2008). The milk volume covered accounted for only a small share of total milk production during the period, measured in billions of pounds. Total milk volumes covered by LGM-Dairy policies have grown but are still relatively small (table 1). The liability has also increased since the product was introduced, and LGM-Dairy policies in 2012 accounted for almost 97 percent of the livestock policies purchased.

How Does LGM-Dairy Work?

The complete instructions for purchasing Livestock Gross Margin-Dairy (LGM-Dairy) products can be found at http://www.rma.usda.gov/livestock/. A short summary is provided here. LGM-Dairy is similar to a "bundled option strategy," purchasing a put option on class III milk futures as well as call options on corn and soybean meal futures. For a short window of time each month, from the last business Friday of each month until the next evening, producers can purchase the insurance for a 10-month period beginning with the second month after the offering month. For example, on the last business Friday of March, coverage can be purchased for milk produced the following May through February.

When insurance is purchased, the producer specifies the amount of milk that he or she produces and plans to insure, as well as expected quantities of corn and soybean meal (SBM) to be fed to reach that level of milk production. There are default values for feed usage, as well as minimums and maximums that can be declared. The prices for milk, corn, and SBM used to determine the margin guarantee are based on a 3-day average of futures market closes for those 3 contracts in the offering month. Following the previous example, on the last business Friday in March, the margin guarantee for the months of May through February would be based on the respective nearby futures closes for each of the 10 contract months for those 3 commodities for the last business Wednesday of March through the last business Friday of March. For example, if on the last business Wednesday-Friday of March, the May class III futures closes were \$16, \$17, and \$18 per hundredweight (cwt), the guarantee price for May would be based on the average, or \$17 per cwt. This same approach is used for corn and soybean meal futures.

As opposed to the guaranteed gross margin, the actual gross margin is later determined by the 3-day average settlement price for class III milk, corn, and SBM on the last 3 trading days of that month. For example, the actual gross margin for June is determined by the prices on the last 3 trading days of June. The indemnity received is the difference in the guaranteed gross margin and the actual gross margin, minus any applicable deductible. Producers can choose to insure their gross margins for any time period within the available 10 months. For example, a producer could insure over the entire 10 available months or choose any number or combination of months to insure. Regardless, indemnities are paid at the end of the insured period whether it is the full 10-month period or a single month.

It is also important to note that indemnities are paid on a gross basis. For example, a producer who insured his or her gross margin in each of the 10 months separately might well receive an indemnity in only 2 or 3 months of the entire period. Conversely, if the same producer chooses to insure the gross margin for the entire 10 months, he or she may not receive an indemnity at all as better months may offset the effect of the weaker months in aggregate (Gould and Cabrera, 2011). This difference is reflected in the premium paid by the producers.

Table 1. Summary statistics related to Livestock Gross Margin-Dairy insurance policies

Reinsurance year	Amount covered	Share of U.S. production	Liability	LGM-Dairy share of all LGMs
	Cwt	*Percent*	*Million $*	*Percent*
2009	401,680	0.02	4.7	10.5
2010	1,872,499	0.10	24.9	31.9
2011	46,172,815	2.37	869.6	81.3
2012	40,504,408	2.03	704.5	96.5
Partial 2013	34,188,752	NA	600.0	89.4

LGM = Livestock Gross Margin. cwt = hundredweight. NA = Not available.

Source: USDA, Economic Research Service using data from Federal Crop Insurance Corporation.

Supply Response and Risk Management

Much of the analysis of agricultural product supply response to risk is based on the work of Chavas and Holt (1990), who modeled corn and soybean acreage decisions by potentially risk-averse producers as a function of price variances and covariances. Chavas and Holt found evidence of wealth effects, implying decreasing absolute risk aversion, some price variance terms were significant, and nontrivial acreage responses were predicted for price support programs.

Bakhshi and Kerr (2009) modified the Chavas and Holt (1990) methodology to isolate insurance effects from market and wealth effects in an application to Canadian field crops. The insurance effects were statistically and economically significant, prompting the authors to conclude that decoupled government payments were production distorting, in conflict with their World Trade Organization (WTO) Green Box status. Lin and Dismukes (2007) also found statistically and economically significant impacts of revenue risk on both soybean acreage and the share of crop acreage planted to soybeans. Not all studies produced evidence of supply response to risk. Luh and Stefanou (1989) found no evidence of supply response to risk based on data from Pennsylvania dairy farms during 1977-84, and Liang et al. (2011) found revenue variance responses with small magnitudes.

Methods Used to Estimate Risk Reduction and Supply Impact

As mentioned in the summary, this study addresses two primary questions. First, how effective is LGM-Dairy as a margin risk management tool? To make that assessment, researchers developed a simulation approach based on examining historical margins and the effects of LGM-Dairy on those margins had the product been in place. Second, assuming risk reduction is achieved, how much of an impact on production (milk supply) would be expected as a result? For information on the approach used to determine the efficacy of LGM-Dairy, see box "Formulating the Model for Gross Margin Risk Analysis" and the results that follow. The supply impacts are discussed in the section "Results From Estimating the Risk Elasticity of Supply."

To summarize, the model underlying the simulation approach is based on risk being defined as the downside squared deviation from a median gross margin. In turn, the gross margin is defined as the milk price minus a feed cost using default feed quantities for corn and soybean meal. Aggregate gross margins are estimated for 13 regions over successive 10-month periods from January 2001 to April 2011, both with and without LGM-Dairy insurance.

The aggregate gross margins constructed for each region using regional mailbox milk prices (net prices received), a regional corn price where available, and national soybean prices represent the "without LGM-Dairy margin insurance" case. The estimates of the gross margin "with the margin insurance" are derived by subtracting insurance premiums and adding indemnities to the aggregate margins estimated without the LGM-Dairy margin insurance.

The analysis relies on two other key assumptions: (1) that LGM-Dairy margin insurance was available each month, and (2) that the milk producers chose to cover 100 percent of their milk production each month over 10 successive 10-month periods. The downside margin risk reduction associated with the assumed LGM-Dairy availability and production coverage is determined for each region.

Formulating the Model for Gross Margin Risk Analysis

The presumption that using LGM-Dairy does reduce milk producers' risk means the analyst(s) must choose how to define "risk."

Often, deviation from some measure of central tendency is selected as that definition. However, milk supply is expected to respond negatively only to downside risk, so only the downside deviations were considered. The other choice to be considered is the measure of central tendency. The mean is often the initial choice, but it may not represent typical outcomes if the distribution of gross margins is heavily skewed. Instead, deviations from the median gross margin were used.

A convention of measuring risk using root mean squared deviations that weight large deviations more heavily than small deviations is applied. Lastly, risk is measured over each 10-month LGM-Dairy contract period, and it is assumed that the producer purchases pooled insurance every month, with each contract representing 10 percent of production during the 10-month contract period (essentially a moving average approach). In this way, 100 percent of each month's production is insured over 10 successive, overlapping LGM-Dairy contract periods. To summarize, risk is defined as root mean squared downside deviations from the median gross margin as defined in LGM-Dairy over a 10-month contract period.

Estimating the Gross Margin Risk for Dairy Producers

Participation in LGM-Dairy requires payment of insurance premiums. The premiums were calculated from the same simulation data used to construct the actual LGM-Dairy premiums. The data are posted at the "Understanding Dairy Markets" website (Gould, 2011) under the section "Underlying Data." The calculation methods follow the guidelines in place since December 17, 2010 (USDA Risk Management Agency, 2010b), and the scenario calculated assumes the default feed ration values of 0.5 bushels (bu) of corn and 4 pounds of soybean meal per hundredweight (cwt) of milk produced, pooled coverage for 10-month contract periods, and a zero deductible.

For each month in each contract period, 5,000 simulated prices of milk, corn, and soybean meal were provided, allowing the calculation of 5,000 simulated "actual" gross margin values. By comparing these values to the gross margin guarantee value for the corresponding contract period, 5,000 simulated indemnities were calculated. The average of the simulated indemnities, plus a 3-percent load, represented the unsubsidized premium used in the analysis.

For reference, premiums were also calculated that included an 18-percent subsidy, which as of June 2013 is associated with a zero deductible (USDA, Risk Management Agency, 2009). Actual polices purchased by farmers spanned a range of deductibles.

After accounting for the premium that is paid with certainty, the remaining impact of participation is indemnity payments, which occur when the Actual Gross Margin falls below the Gross Margin Guarantee. Historical data from January 2001 through April 2011 were used to calculate margins and indemnities from January 2002 to May 2010 as if LGM-Dairy had been in place during the entire period. The calculations follow the default feed coefficients and other contract specifications used in 2011 and do not use prior calculation methods, such as the basis adjustments for milk and corn that were used before July 2009. A zero deductible was assumed so the maximum risk reduction could be evaluated. With no deductible, the Gross Margin Guarantee is equal to the Expected Gross Margin.

The Gross Margin Guarantee is the sum over each 10-month contract period of monthly gross margins calculated from appropriately deferred futures prices for class III milk, corn, and soybean meal. The historical deferred futures prices are conveniently collected online in the "Underlying Data" section of Gould (2011). Gross margin guarantee calculations were performed using assumed parameters for the milk quantity insured (1 cwt), a zero deductible, and default feed coefficients of 0.5 bu/cwt for corn and 0.002 ton/cwt for soybean meal.

The Actual Gross Margin was calculated using the same parameters and formulas, again using historical data provided online by Gould (2011). In this case, "actual" prices represent the average of the final 3 days of milk, corn, or soybean meal futures settlement prices before expiration. For months with no futures contract, a weighted average of surrounding months is used instead. At this point, indemnities were calculated.

The term "Actual Gross Margin," while necessary for the calculation of indemnities, does not describe the gross margin realized by an individual producer in a specific location. The effectiveness of LGM-Dairy at reducing gross margin risk might vary widely across space, especially in locations or times where the class III prices represented less than the majority of the all-milk price. For clarity, a producer's local gross margin is referred to here as "Realized Gross Margin."

The most feasible way to approximate Realized Gross Margin is to use region-specific mailbox milk prices, available online from Gould (2011), and State-level monthly average corn prices where possible (USDA, National Agricultural Statistics Service, various years). National average soybean meal prices are used, as the 10 cash markets for which USDA's Agricultural Marketing Service collects soybean meal prices align poorly with the regions evaluated in the present analysis.

Just as producers who hedge in futures and options markets face basis risk, participants in LGM-Dairy face an analogous risk in that changes in realized gross margins may not be highly correlated with indemnity payments. There are two causes of potentially low correlation, the first being that national-level class III milk and corn price changes are imperfectly correlated with State and regional-level price variation. The second cause is that indemnities do not necessarily occur when absolute gross margin levels fall; indemnities occur when gross margins fall from higher expected levels during the life of an insurance contract. One can receive indemnities when gross margins are high but not as high as expected, and one can fail to receive indemnities when gross margins are very low. This is the nature of using futures and options markets as elements of the insurance product.

Historical monthly mailbox milk prices were gathered for the following available regions: Northwest, California, New Mexico, Texas, Minnesota, Wisconsin, Illinois, southern Missouri, Michigan, Ohio, Kentucky, Florida, and New England. State-level monthly-average corn prices were available for a subset of these regions: Texas, Minnesota, Wisconsin, Illinois, Missouri, Michigan, Ohio, and Kentucky. Regional Realized Gross Margins were approximated using as much localized data as were available, and these represent outcomes without participation in LGM-Dairy. Indemnities were added to the Realized Gross Margins, and premiums were deducted, to calculate the net Realized Gross Margins with participation in LGM-Dairy.

Regional average Realized Gross Margins, with and without participation, were next calculated over the period January 2002 to May 2010 under two scenarios: no premium subsidy and an 18-percent premium subsidy. These results are useful in estimating supply response to gross margin levels. Similarly, root mean-squared deviation from median outcomes was calculated for each region, and each premium subsidy level, with and without participation in LGM-Dairy.

In summary, the gross margin was calculated with and without insurance (including both indemnities received and premiums paid), the measure of risk was calculated for both scenarios, and the percentage reduction in risk attributable to LGM-Dairy was calculated. Using risk response elasticities from the literature and original dairy supply modeling results, an associated percentage change in supply was attributed to LGM-Dairy's impact on margin risk levels.

In addition to affecting risk, LGM-Dairy can also affect the mean level of gross margins, either because of loaded or subsidized premiums. The percentage change in mean gross margins over each 10-month contract period was calculated and applied to price elasticity of supply estimates to predict the percentage change in supply due to policy-induced changes in margin levels.

Results of Gross Margin Risk Analysis

Findings suggest that LGM-Dairy was effective in reducing the risk levels of dairy producers. The results reported in this section are based on the model framework and assumptions described in the previous section. The results that any individual producer might expect would be adjusted to reflect actual prices and feeding decisions. The root mean-squared deviation from the median margin was found to be considerably smaller when LGM-Dairy was used than when it was not used. The results are quite robust, as a considerable reduction in risk level was found for each of the regions analyzed, ranging from 28 percent in Minnesota to 39 percent in Florida (table 2). Moderate risk-reduction levels help account for the popularity of the program.

Findings reveal that risk reduction was indeed achieved in each of the 13 regions included in the analysis. Previous analysis of the Dairy Options Pilot Program found risk reduction across regions, but the reduction was greater in those regions where cheese manufacturing (class III use) was higher (Maynard et al., 2005). The results of this analysis would suggest that basis risk across regions is perhaps less of a concern than previously thought because similar risk reduction rates were found in all regions. Factors that may account for this effect include the index mechanism itself, the inclusion of feed costs in LGM-Dairy, and the assumed price risk strategy of insuring 100 percent of monthly output over 10 successive insurance contracts.

Table 2. Risk reduction associated with LGM-Dairy (January 2002 - May 2010)

Region	Risk level* without LGM-Dairy	Risk level* with LGM-Dairy	Reduction in risk
	Dollars per cwt	*Dollars per cwt*	*Percent*
Northwest	21.06	13.49	36
California	20.76	13.23	36
New Mexico	19.16	12.65	32
Western Texas	19.59	12.65	35
Minnesota	21.36	15.29	28
Wisconsin	22.03	15.71	29
Illinois	23.49	15.92	32
Southern Missouri	21.56	14.50	33
Michigan	23.61	15.18	36
Ohio	22.75	15.23	33
Appalachian	20.72	14.31	31
Florida	21.92	13.45	39
New England	25.35	17.58	31

*Risk level is defined as the root mean-squared downside deviations from the median gross margin. Insurance assumptions include zero deductible and 18 percent premium subsidy. Risk levels are the aggregate over the relevant 10-month period.
cwt = hundredweight. LGM = Livestock Gross Margin.
Source: USDA, Economic Research Service, based on model results.

While consideration of risk reduction levels is certainly important, so, too, are evaluations of realized margins with and without LGM-Dairy. LGM-Dairy premiums contain a 3-percent catastrophic reserve load (see glossary), which is expected to result in lower average margins for those producers using LGM-Dairy. Premium subsidies, however, are expected to more than offset the impact of the reserve load, possibly producing a net gain in average margin. Thus, supply growth could occur from both lower risk and higher average returns. However, as shown in table 3, the change in average margin ranged from -1 to 2 percent, with virtually no change when locations were pooled.

Results reported in tables 2 and 3 both assume a zero deductible, which is associated with an 18-percent premium subsidy. Higher deductibles are generally associated with higher percentages in subsidy levels (the LGM-Dairy subsidy level for the highest deductible was raised after this analysis was completed to a maximum of 50 percent). Still, the analysis also examines risk reduction and gross margin impacts for alternative levels of deductible and their associated premium subsidy, as choosing among the deductible amounts is an option available to producers. Table 4 reports the risk reduction and gross margin results averaged across all regions. The greatest risk reduction and lowest margin penalty occur simultaneously when the deductible level is zero, suggesting there is little incentive for producers to choose higher deductible levels, despite the increased premium subsidy.

RESULTS FROM ESTIMATING THE RISK ELASTICITY OF SUPPLY

Estimating the milk supply response to a risk-reducing policy tool such as LGM-Dairy requires two major steps: estimating risk reduction (described in the preceding section) and estimating marginal supply response to a unit of risk reduction (a risk elasticity). Chavas and Holt (1990) and subsequent related studies (e.g., Lin and Dismukes, 2007) estimated both price and risk elasticities of supply. In this study, the milk supply component of the ERS forecasting model documented in Mosheim (2012) was used to estimate supply response. As in the previous section, the focus here is on the results of the modeling effort rather than the details (see appendix).

The impact of risk on production, defined as production per cow (PPC), was significant at roughly the 90-percent level but very small at -0.006. This result implies that a 10-percent reduction in risk would lead to an increase in production of less than 1 (.6) percent. This small, marginally significant finding is consistent with the findings of Luh and Stefanou (1989). Given the range of risk elasticity estimates from the literature and the results of the ERS modeling effort, a sensitivity table was deemed an appropriate way to present supply impact estimates. Note, also, that the risk elasticity of-0.006 estimated using the ERS model, which falls below the low end of this analysis, would suggest risk impacts of approximately zero (table 5). Based on the estimates in table 5, short-term milk production impacts associated with use of LGM-Dairy are small. Even at the highest risk elasticity levels, supply impacts never exceed 4 percent and the estimated risk elasticity of -0.006 derived from the ERS modeling effort is consistent with supply impacts well below 1 percent. Note, also, that the responses indicated in table 5 assume full participation in LGM-Dairy. Actual participation has been much lower, suggesting correspondingly lower supply impacts. Further, it is worth noting that longrun supply response would likely be even smaller.

Table 3. Average realized margin by region (January 2002-May 2010)

Region	Aggregate margin without LGM-Dairy	Aggregate margin with LGM-Dairy	Change in margin
	Dollars per cwt	*Dollars per cwt*	*Percent*
Northwest	121.62	124.27	2
California	115.69	118.15	2
New Mexico	114.73	117.03	2
Western Texas	122.59	124.70	2
Minnesota	137.98	137.95	0
Wisconsin	138.46	136.77	-1
Illinois	138.07	136.35	-1
Southern Missouri	136.15	134.48	-1
Michigan	134.99	133.31	-1
Ohio	140.99	139.22	-1
Appalachian	147.21	145.50	-1
Florida	168.01	166.26	-1
New England	146.40	144.24	-1

* Insurance assumptions include zero deductible and 18 percent premium subsidy. Aggregate margins are reported over the relevant 10-month period.

cwt = hundredweight. LGM = Livestock Gross Margin.

Source: USDA, Economic Research Service, based on model results.

Table 4. Risk reduction and gross margin impacts of Livestock Gross Margin-Dairy participation under various deductible subsidy combinations

Deductible	Corresponding premium subsidy	Risk reduction	Change in average gross margin
Dollars per cwt	*Percent*		
0.00	18	33	0
0.20	21	31	-1
0.40	25	28	-1
0.60	31	25	-1
0.80	38	23	-1
1.00	48	20	-1
1.50	50	14	-1
2.00	50	9	-2

cwt = hundredweight.

Source: USDA, Economic Research Service, based on model results.

Table 5. Expected supply response given risk-reduction level and elasticities

Region	**Reduction in risk**	**Supply impact: E_{risk}=-0.10**	***Supply impact: E_{risk} = -0.05***	***Supply impact: E_{risk} = -0.025***
		Percent		
Northwest	37	3.66	1.83	0.91
California	37	3.72	1.86	0.93
New Mexico	34	3.36	1.68	0.84
Western Texas	36	3.59	1.80	0.90
Minnesota	28	2.84	1.42	0.71
Wisconsin	29	2.89	1.44	0.72
Illinois	33	3.29	1.64	0.82
Southern Missouri	35	3.52	1.76	0.88
Michigan	36	3.58	1.79	0.89
Ohio	35	3.46	1.73	0.87
Appalachian	33	3.27	1.64	0.82
Florida	39	3.86	1.93	0.96
New England	31	3.12	1.56	0.78

Source: USDA, Economic Research Service, based on model results.

Table 5 shows the supply impact attributable to risk only. The other source of supply impact is changes in average gross margin, with premium subsidies reducing the negative supply impact of LGM-Dairy premiums. Given the nationwide average-margin change of zero percent associated with the zero-deductible alternative shown in table 4, average-margin changes are expected to be a negligible source of aggregate supply effects, so most of the supply effects should come from risk reduction.

CONCLUSIONS, IMPLICATIONS, AND AREAS FOR FURTHER RESEARCH

This study examines the implications of LGM-Dairy. From the perspective of the dairy producer, LGM-Dairy appears to offer an opportunity to reduce downside gross margin risk by approximately one-third when measured as downside deviations from the median gross margin over a 10-

month period. Evidence from the present analysis and prior literature also suggests that LGM-Dairy may offer risk reduction exceeding that provided by hedging with dairy options alone. The findings from this study reveal that reductions in risk are similar across regions, a result that alleviates potential concerns that programs of this type are not as effective for producers in regions who supply a large share of their output to fluid markets (high class I use areas).

Other analyses of programs similar to LGM-Dairy have added to the knowledge base for evaluating such programs in the future. These include works by Bozic et al. (2012) and Stephenson (2012), which examine recent legislative proposals for a voluntary dairy margin insurance program for U.S. milk producers. Note that these legislative programs are very different from the existing LGM-Dairy program. As the "safety nets" for agricultural producers are further considered, even more information about the LGM-Dairy program will likely be requested.

For risk reduction, a clear advantage of LGM-Dairy over using price futures and options is the ability to protect a margin rather than a single output price. Unlike class III dairy options, which can only be triggered when milk prices fall, indemnities can be received on LGM-Dairy policies when milk prices fall, feed prices rise, or some combination of the two. Further research in this area might involve analyzing the risk-reduction effects of LGM-Dairy in situations where corn and soybean meal are not the primary feeds being purchased (such producers can still purchase LGM-Dairy but must convert the feeds they are using to corn/soybean feed equivalents), where only minimal amounts of purchased feeds are used, or where the dairy operation relies largely on home-grown feeds. An additional "basis risk" would exist for grazing operations or for operations that purchase and feed large quantities of hay or other feeds beyond corn and soybean meal.

While one might assume that reduced risk would have supply effects, the literature reviewed was not especially robust as to the magnitude of this effect. Results of direct risk modeling using quarterly milk price and production data suggest a shortrun supply increase of less than 1 percent if all producers participated in LGM-Dairy. Given that actual participation has been much lower, partly due to limited funding and program availability, the estimated supply impact over the last few years would be even lower.

GLOSSARY

Actuarial soundness: An insurance term describing a situation in which indemnities paid, on average, are equal to total premiums collected.

Adverse selection: A situation in which an insured has more information about his or her risk of loss than does the insurance provider and is better able to determine the soundness of premium rates.

Basis: The difference between a specific futures price and a specific cash price for the same or related commodity.

Basis risk: The risk associated with an unexpected widening or narrowing of the basis between the time a hedging position is established and the time that it is lifted.

Bundled option insurance: In the case of LGM-Dairy, similar to buying both call options on multiple feeds to limit higher feed costs and a put option to set a floor on milk prices.

Catastrophic reserve load: An adjustment intended to account for infrequent, severe events that are not fully captured in the base premium rate.

Deductible: The portion of the expected gross margin that producers elect not to insure. Per hundredweight (cwt) deductible amounts range from zero to $2.00 per cwt in 10 cent increments. The deductible equals the selected per cwt deductible times the sum of target marketings across all months of the insurance period.

Downside deiviation from median: In the case of this analysis, the difference between the median risk level and actual risk level that was used to evaluate the risk reduction effects of LGM-Dairy.

Forward contract: An agreement between two parties (such as you and someone who buys your products) that calls for delivery of, and payment for, a specified quality and quantity of a commodity (such as a particular crop) at a specified future date. The price may be agreed upon in advance or determined by formula at the time of delivery or other point in time.

Futures contract: An agreement priced and entered into on an exchange to trade at a specified future time a commodity, or other asset, with specified attributes (or in the case of cash settlement, an equivalent amount of money).

Gross margin guarantee: The expected total gross margin minus the deductible for an insurance period.

Hedging: Buying or selling in a futures or options market intended as a temporary substitute for the later actual sale or purchase of a commodity. Its aim is to protect against adverse price movements prior to the actual transaction.

Indemnity: The compensation received by an individual for qualifying losses paid under an insurance policy. The indemnity compensates for losses that exceed the deductible up to the level of the insurance guarantee.

Moral hazard: The ability of an insured to increase his or her expected indemnity by actions taken after buying the insurance.

Options contract: A contract that gives the holder the right, without obligation, to buy or sell a futures contract at a specific price within a specified period of time, regardless of the market price of the futures.

Premium: An amount of money paid to secure risk protection. Option buyers pay a premium to option sellers for an options contract. Similarly, the purchaser of an insurance policy pays a premium to obtain coverage.

Reinsurance: A method of transferring some of an insurer's risk to other parties. In the case of LGM-Dairy, USDA's Federal Crop Insurance Corporation shares the risk of loss with each private insurance company delivering policies to producers. Private reinsurance also exists, in which case, a private reinsurer assumes responsibility for a share of the risk in return for a share of the premiums.

Risk: Uncertainty about outcomes that are not equally desirable. Risk may involve the probability of making (or losing) money, harm to human health, negative effects on resources (such as credit), or other types of events that affect welfare.

Target marketings: The quantities of milk producers elect to insure in each month of the insurance period.

APPENDIX AGGREGATE SUPPLY RESPONSE MODELING

Underlying the model is a shortrun, or restricted, aggregate profit function specification that in the most general terms is defined as:

$$\pi^* = \max_y \{p^* y - VC(y, w^*) - K - \tfrac{1}{2}\lambda y^2 \sigma^2\} \qquad (1)$$

where: π^* is expected profit;

p^*, σ^2 are ex ante mean and variance of output price;
y is output;
w^* is a vector of ex ante input prices;
$VC(y, w^*)$ is variable cost;

K is capital; and

λ is the coefficient of risk aversion.

First-order conditions imply that marginal cost equals marginal revenue. Since risk was modeled in equation (1), this first-order condition will mean that the profit-maximizing firm will produce a lower level of output in a riskier environment than otherwise,

$$p^* - MC\,(y, w^*, K) - \lambda y \sigma^2 = 0. \tag{2}$$

From this first-order condition, a shortrun output supply function takes the form,

$$y^{SR} = y(p^*, w^*, \sigma^2, K). \tag{3}$$

A specification of the shortrun function presented in Moschini and Hennessy (2001) is:

$$y_t = \alpha_o + x_t^{'}\beta + \delta_1\mu_t + \delta_2\sigma_t^2 + \varepsilon_t\,, \tag{4}$$

where

y_t is supply;

$x_t^{'}$ is a vector of other variables affecting decisions;

μ_t is the ex ante output price mean;

σ_t^2 is the ex ante output price variance;

α_o, β (a vector), δ_1, δ_2 are parameters to be estimated;

$\varepsilon_t \sim N(0,\sigma^2)$; and t indexes the observations.

This functional form is employed to model the uncertainty inherent in the variation of expectations of future milk prices, as compared with the first model that uses only actual price data for analysis.

The empirical specification of the statistical supply model consists of three equations, with the dependent variables being production per cow, cow numbers, and the all-milk price. These equations are not simultaneous but have a recursive specification. The first "run" of the model yields empirical estimates of the supply response to both price and expected price, which can then be applied to the estimated policy-induced magnitudes of price and risk changes to estimate total supply impacts of LGM-Dairy. Two specifications are presented. One is identified with the cash market, which serves as a baseline, and the second seeks to incorporate risk to various degrees. The "first-run" model provides an empirical supply function and is used to derive an elasticity of supply.

Two situations are examined, the first being the cash market situation where the producer uses only prices to make decisions that will have an economic impact sometime in the future for his or her enterprise once milk is sold in the market. This can be thought of as a naïve model where producers make decisions based on current prices. Second, a model is developed that integrates expected price and its variance (at this point only the output price is considered), with expected prices defined by deferred futures prices. The second model allows estimation of a risk elasticity of supply. The results of the estimation of the two models are presented in tables A2 and A3.

Data Sources for Aggregate Milk Supply Modeling

The data for estimating the statistical supply models came from several sources. The basic information on herd size (COWS), milk per cow (PPC), and the all-milk price are from USDA's National Agricultural Statistics Service (NASS). Supply is defined as cow numbers multiplied by output per cow in the two models. The all-milk price, AMP, is viewed as the incentive price to dairy farmers. In the second model, a futures price for AMP, denoted by AMP^*, is needed, which is derived in turn from class III and class IV futures information. These future class prices were converted to real values by employing the Gross Domestic Product (GDP) deflator as forecasted by USDA's Interagency Commodity Estimates Committee (ICEC). The other variables include slaughter cow price (SCP), feed cost price (FC), a dummy variable for the spring quarter (DQ2), and time trends.

A regression of the AMP price on class III and class IV prices was employed to convert the future class prices into AMP^* projections. The last 200 future closing prices available for a particular contract that expires in the quarter being projected were employed for quarters I 2006 through IV 2010. Expected mean and variance were calculated from these announcements. AMP data are available in *Agricultural Prices*, a monthly USDA publication. Feed costs and slaughter cow prices are either historical data or forecasts obtained from ICEC, which forecasts eight quarters for FC, SCP, and the GDP deflator, with the number of observations employed as inputs for the models denoted in parentheses. All prices are deflated by the GDP deflator. Summary statistics for key variables are shown in table A1.

Specification and Estimation of the Supply Models

The following models intend to capture the effect of risk on production decisions by dairy farmers. First, in what is called the "market model," farmers use published prices to make production decisions, a behavior inherently riskier than that captured in the second model, where farmers employ the futures market. The latter estimation, the "futures market model," is an augmented version of the cash market model, and it employs the expected mean and variance of future all-milk price. The models follow:

Table A1. Variable and sample means (standard deviation) (number of observations in squared parentheses)

Variable (quarters) (Units)	Mean	Std. Dev.	Variable (quarters) (Units/Definition)	Mean	Std. Dev.
COWS [20] (Thousands)	9,192	83	FC[28](Feed cost price, 2005 $)	7.12	1.34
PPC [20] (Pounds/Cow)	5,111	141	DQ2 [28] Spring	0.21	0.42
AMP* [20] (All-milk price "Futures," 2005 $)	14.61	1.00	t [28] (time trend)	15	8.23
AMP [20] (All-milk price, 2005 $)	14.77	2.91	t2[28] (time trend2)	276	246
SCP [28] (Slaughter cow price, 2005 $)	52.71	8.08	GDP deflator [28] (2005 = 1)	1.09	0.04

Source: USDA, Economic Research Service, based on model results.

I. Cash Market Model

$$\ln COWS_t = \chi_0 + \chi_1 \ln COWS_{t-1} + \chi_2 \ln COWS_{t-2} + \chi_3 \ln AMP_{t-2} + \chi_4 \ln FC_{t-2} + \chi_5 t^2 \quad (5)$$

$$\ln PPC_t = \theta_0 + \theta_1 \ln PPC_{t-2} + \theta_2 \ln AMP_{t-2} + \theta_3 \ln FC_{t-2} + \theta_4 DQ2 + \theta_5 t \quad (6)$$

$$\ln AMP_t = \alpha_0 + \alpha_1 \ln COWS_{t-2} + \alpha_2 \ln PPC_{t-1} + \alpha_3 \ln SCP_{t-1} + \alpha_4 \ln FC_{t-1} \quad (7)$$

Total milk production is a reflection of the number of cows and their productivity (defined as production per cow). As in any production system, inputs are transformed into outputs using a given technology, in a given environment, and within a specific set of constraints. Two important factors were modeled explicitly in this set of equations—feed costs and the all-milk price.

Equation (5) represents cow inventory. The number of cows is expected to be a function of lagged cow inventory (COWS), lagged price of milk (AMP), and lagged feed costs (FC). Real values for these price and cost variables are determined by dividing current magnitudes by the GDP deflator. Real values for the price variable symbols do not change by dividing by GDP deflator. In the case of the AMP variable, direct payments, such as those from the Milk Income Loss Contract (MILC) program, were added before converting the all-milk price variable to a real value to model the farmer's incentive or effective price. The variable AMP is expected to have a positive lagged effect on the number of cows. It can be expected that as FC (approximately 60 percent of dairy farm operating costs according to USDA's Agricultural Resource Management Survey) increases, all other things being equal, there will be a lagged decrease in herd size. Both AMP and FC effects are related to biological lags, culling, and replacements apart from the other variables. In the double logarithm specification of the equations, the coefficients can be interpreted as elasticities.

Equation (6) captures the combined effect of new technologies and factors like scale economies on cow productivity. Generally, larger operations have higher cow productivity, represented by the variable production per cow (PPC). In equation (7), total farm milk processed is decomposed into PPC and number of cows (COWS), and their respective price (AMP) elasticities show the effect of productivity on AMP as being very inelastic and that of the number of cows as very elastic (i.e., an increase in cow productivity has a larger downward effect on AMP than a comparable percentage change increase in the dairy herd, all other things equal). Feed price increases, ceteris paribus, decrease supply (less milk supply at the same milk price—leftward supply shift), requiring a higher AMP in this equation to equilibrate supply with derived demand.

II. Futures Market Model

$$\ln COWS_t = \chi_o + \chi_1 \ln COWS_{t-1} + \chi_2 \ln COWS_{t-2} + \chi_3 \ln AMP^*_{t-2} + \chi_4 \ln FC_{t-2} + \chi_5 t^2 + \chi_6 \sigma_{\ln AMP^*_{t-1}} \tag{8}$$

$$\ln PPC_t = \theta_o + \theta_1 \ln PPC_{t-2} + \theta_2 \ln AMP^*_{t-2} + \theta_3 \ln FC_{t-2} + \theta_4 DQ2 + \theta_5 t + \theta_6 \sigma_{\ln AMP^*_{t-2}} \tag{9}$$

$$\ln AMP^*_t = \alpha_o + \alpha_1 \ln COWS_{t-2} + \alpha_2 \ln PPC_{t-1} + \alpha_3 \ln SCP_{t-1} + \alpha_4 \ln FC_{t-1} + \alpha_5 \sigma_{\ln AMP^*_{t-3}} \tag{10}$$

$$\sigma_{\ln AMP^*_t} = v_o + v_1 \sigma_{\ln AMP^*_{t-1}} + v_2 t + v_3 \sigma_{\ln AMP^*_{t-2}} + v_4 t^2 \tag{11}$$

The expectation, employing Chavas (2004, chapter 8), is that production under uncertainty will have a negative effect on cow numbers and production per cow and a positive effect on the all-milk price. That is, the elasticity of supply under uncertainty will be steeper (less responsive) than that where the underlying price information is more certain. Two estimations are conducted to determine empirically the quantitative effect of risk. From the equations (5) through (11) estimated for both the cash market model (presented in table A2) and the futures market model (presented in table A3), predicted supply functions are derived for the cash (figure A1) and futures (figure A2) markets.

These models are estimated using nonlinear three-stage least squares. The first model uses NASS prices and the second employs both the expected mean and variance of futures prices. Coefficients with a t-value greater than 1 were considered for specification purposes to decrease type II errors as suggested in Kennedy (2008, p. 90). Also, on theoretical grounds, some variables with t-values less than 1 were considered.

From the above model, predicted results on the number of cows and production per cow can be employed to derive a supply function, and two sets of prices are used to make inferences about risk: the all-milk price from NASS (the cash market) and the derived all-milk futures price, both described above. The most essential result from the two figures is the derived price elasticity of supply that declines as more uncertainty is assumed by producers, an approximation to what Chavas (2004, p. 99) calls the marginal risk premium.

Table A2. Cash market model (t statistics in parentheses)

Parameter	Estimate	Parameter	Estimate	Parameter	Estimate
χ_0	5.32 (4.50)	θ_0	12.92 (13.09)	α_0	158.30 (2.61)
χ_1	0.64 (3.16)	θ_1	-0.52 (-4.51)	α_1	-15.02 (-2.47)
χ_2	-0.23 (-1.21)	θ_2	0.05 (2.88)	α_2	-2.59 (-2.13)
χ_3	0.02 (3.95)	θ_3	-0.06 (-3.80)	α_3	0.60 (1.89)
χ_4	0.01 (1.37)	θ_4	0.02 (2.76)	α_4	0.68 (2.02)
χ_5	-2.00E-5 (-2.40)	θ_5	0.007 (10.87)		

Source: USDA, Economic Research Service, based on model results.

Table A3. Futures market model, expected all milk price mean and variance (t statistics in parentheses)

Parameter	Estimate	Parameter	Estimate	Parameter	Estimate
χ_0	0.00 (0.00)	θ_0	13.60 (18.75)	α_0	-17.77 (-0.50)
χ_1	1.00 (5.88)	θ_1	-0.64 (-7.48)	α_1	2.28 (0.63)
χ_2	-0.02 (-0.07)	θ_2	0.17 (3.34)	α_2	-0.12 (-0.21)
χ_3	0.10 (2.63)	θ_3	-0.07 (-2.50)	α_3	0.12 (0.64)
χ_4	-0.04 (-1.58)	θ_4	0.02 (5.88)	α_4	0.08 (0.34)
χ_5	0.000004 (0.25)	θ_5	0.007 (12.51)	α_5	-0.04 (-2.02)
χ_6	-0.0008 (-0.82)	θ_6	-0.006 (-1.91)	v_o	-6.58 (-2.71)
		v_1	0.10 (0.47)	v_2	-0.25 (-0.89)
		v_2	0.76 (2.32)	v_4	-0.03 (-2.59)

Source: USDA, Economic Research Service, based on model results.

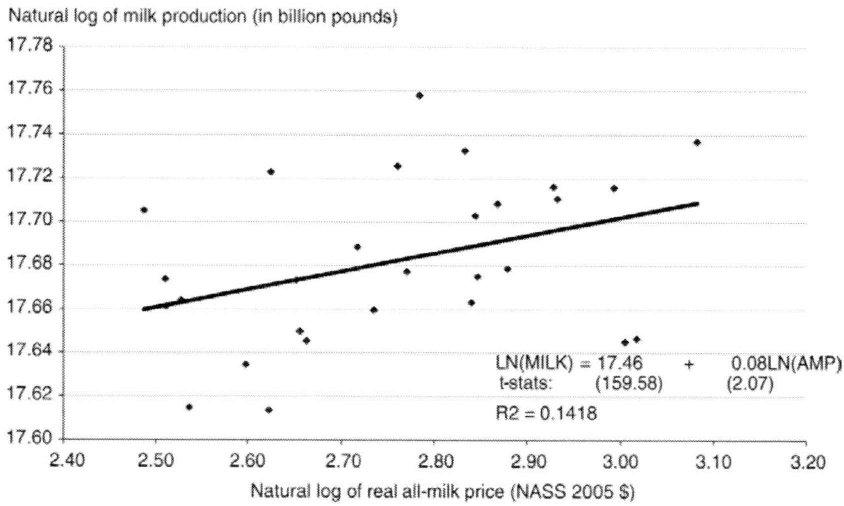

NASS = USDA, National Agricultural Statistics Service.
Source: USDA, Economic Research Service, based on model results.

Figure A1. Supply function in cash market.

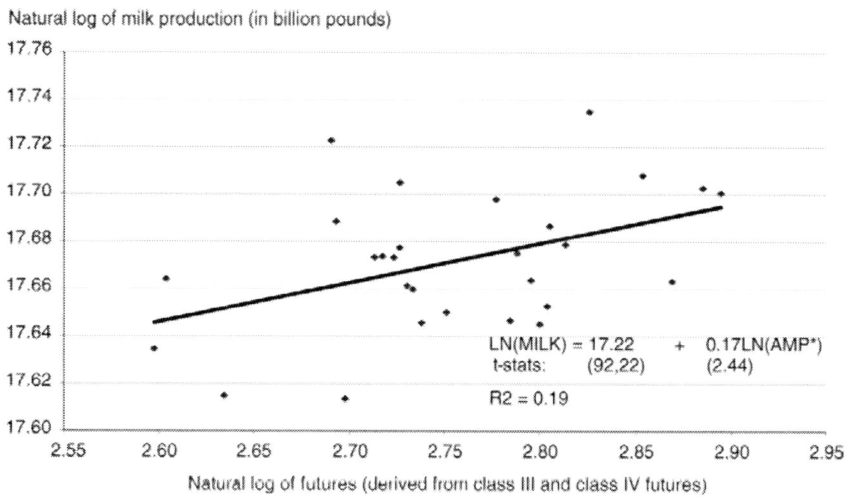

NASS = USDA, National Agricultural Statistics Service.
Source: USDA, Economic Research Service, based on model results.

Figure A2. Supply curve under uncertainty (use of futures market).

REFERENCES

Bakhshi, S. & Kerr. W. A. (2009). "Is There Supply Distortion in the Green Box? An Acreage Response Approach." CATPRN Working Paper 2009-04, Canadian Agricultural Trade Policy Research Network, Ottawa, Canada.

Barnett, B. J. (2004). "Agricultural Index Insurance Products: Strengths and Limitations." Presented at the 2004 Agricultural Outlook Forum. February 19.

Bozic, M., Newton, J., Thraen, C. S. & Gould, B. W. (2012). "Mean-Reversion in Income Over Feed Cost Margins: Evidence and Implications for Managing Margin Risk by U.S. Dairy Producers." *Journal of Dairy Science*, *95*(12), 7417-7428.

Buschena, D. & McNew, K. (2005). "An Exploration of Market Pricing Efficiency: The Dairy Options Pilot Program." Presented at the American Agricultural Economics Association Annual Meeting. Providence, RI. July.

Chavas, J. P. (2004). *Risk Analysis in Theory and Practice*. Elsevier: San Diego, CA. Chavas, J.P. & M.T. Holt. (1990). "Acreage Decisions Under Risk: The Case of Corn and Soybeans." *American Journal of Agricultural Economics*, *72*, 529-38.

Food and Agricultural Policy Research Institute (FAPRI). (2012). "Dairy Policy Issues for the 2012 Farm Bill." Dairy Policy Analysis Alliance.

Gould, B. (2011). "Understanding Dairy Markets: Livestock Gross Margin Insurance for Dairy Cattle." Accessed various dates at http://future.aae.wisc.edu/lgm_dairy.html. Agricultural and Applied Economics, University of Wisconsin, Madison, WI.

Gould, B. & Cabrera, V. (2011). "USDA's Livestock Gross Margin Insurance for Dairy: What It Is and How Can It Be Used for Risk Management." New York State Department of Agriculture and Markets, Albany, NY. September 28.

Harwood, J., Heifner, R., Koble, K., Perry, J. & Somwaru, A. (1999). *Managing Risk in Farming: Concepts, Research, and Analysis*. U.S. Department of Agriculture, Economic Research Service. Agricultural Economic Report No. 774. March.

Ibendahl, G., Maynard, L. J. & Branstetter, A. (2002). "Measuring the Perceived Effectiveness of Training for the Dairy Option Pilot Program." *Journal of Extension.* Volume 40, Number 6. Kennedy, P. (2008). *A*

Guide to Econometrics, 6th Edition. Blackwell Publishing: Oxford, England.

Liang, Y., Miller, J. C., Harri, A. & Coble, K. H. (2011). "Crop Supply Response Under Risk: Impacts of Emerging Issues on Southeastern U.S. Agriculture." *Journal of Agricultural and Applied Economics*, 43, 2, 181-94.

Lin, W. & Dismukes, R. (2007). "Supply Response Under Risk: Implications for Counter-Cyclical Payments' Production Impact." *Review of Agricultural Economics*, *29*, 1, 64-86.

Luh, Y. -H. & Stefanou, S. E. (1989). "Dairy Supply and Factor Demand Response to Output Price Risk: An Econometric Assessment." *Northeastern Journal of Agricultural and Resource Economics*, *18*, 2, 103-08.

Maynard, L. J., Wolf, C. & Gearhardt, M. (2005). "Can Futures and Options Markets Hold the Milk Price Safety Net? Policy Conflicts and Market Failures in Dairy Hedging." *Review of Agricultural Economics*, *27*(Summer 2005), 273-286.

Moschini, G. & Hennessy, D. A. (2001). "Uncertainty, Risk Aversion, and Risk Management for Agricultural Producers." Chapter 2 in *Handbook of Agricultural Economics*, *Volume 1A*. Edited by B.L. Gardner and G.C. Rausser. Elsevier: Amsterdam, Netherlands.

Mosheim, R. (2012). *A Quarterly Econometric Model for Short-Term Forecasting in the U.S. Dairy Industry*. U.S. Department of Agriculture, Economic Research Service. Technical Bulletin No. 1932. January.

Stephenson, M. (2012). *The Impacts on Dairy Farmers and Milk Markets of a Stand-Alone Dairy Producer Margin Insurance Program*. University of Wisconsin.

Stoll, H. R. & Whaley, R. E. (1993). *Futures and Options: Theory and Applications*. South-Western Publishing Co.: Cincinnati, OH.

U.S. Department of Agriculture, National Agricultural Statistics Service. (various years). "Agricultural Prices Summary." Pr 1-3.

U.S. Department of Agriculture, Risk Management Agency. (2009). "Livestock Gross Margin for Dairy Insurance Policy: Step by Step Instructions to Calculate Premium." Accessed May 2012 at www.rma.usda.gov/policies/2010/lgm/10lgmdairypremcalc.pdf. July.

U.S. Department of Agriculture, Risk Management Agency. (2010a). "Livestock Gross Margin Insurance: Dairy Cattle." Program Aid No. 2021. Accessed May 2012 at www.rma.usda.gov/ pubs/rme/lgmdairy.pdf. November.

U.S. Department of Agriculture, Risk Management Agency. (2010b). "Livestock Gross Margin for Dairy Insurance Policy: Step by Step Instructions to Calculate Premium." Accessed September 2011 at http://future.aae.wisc.edu/lgm-dairy/rma_material/11lgmdairypremcalc.pdf, Washington, DC. October.

U.S. Department of Agriculture, Risk Management Agency. (2011). "Demand Strong for the Livestock Gross Margin Dairy Program." News Release, Spokane Regional Office, Spokane Valley, WA. December 15.

Wright, P. J. (2012). "Testimony of Patrick Joseph "Joe Wright," Southeast Milk, Inc., Before the House of Representatives, Committee on Agriculture, Subcommittee on Livestock, Dairy, and Poultry." House Committee on Agriculture, April 26.

End Notes

[1] The analysis presented in this study preceded passage of the Agricultural Act of 2014.

[2] Contract specifications can be found at: www.cmegroup.com.

INDEX

D

E

F

G

H

Q

R

S

T

U

V

W

Y